Implementation of biological filtration for the treatment of cyanobacterial metabolites

Emma Sawade, Lionel Ho, Daniel Hoefel, and Gayle Newcombe

Australian Water Quality Centre, South Australian Water Corporation

AF794561

DISCLAIMER

Water Research Australia Limited (WaterRA), is a not for profit company funded by the Australian water industry.

WaterRA and individual contributors are not responsible for the outcomes of any actions taken on the basis of information in this research report, nor for any errors and omissions.

WaterRA and individual contributors disclaim all and any liability to any person in respect of anything, and the consequences of anything, done or omitted to be done by a person in reliance upon the whole or any part of this research report.

This research report does not purport to be a comprehensive statement and analysis of its subject matter, and if further expert advice is required, the services of a competent professional should be sought.

© Water Research Australia Limited 2014

Location:

WaterRA Head Office
Level 3, 250 Victoria Square, Adelaide SA 5000
Postal Address:

GPO BOX 1751, Adelaide SA 5001
For more information about WaterRA visit the website www.waterra.com.au

Implementation of biological filtration for the treatment of cyanobacterial metabolites

Final Report Project 1021

ISBN 978-1-921732-24-9

Co-published by IWA Publishing
Alliance House, 12 Caxton Street, London SW1H 0QS, UK
Tel. +44 (0)20 7654 5500, Fax +44 (0)20 7654 5555
publications@iwap.co.uk
iwapublishing.com

ISBN: 9781780407531 (Paperback)
ISBN: 9781780407548 (eBook)

FOREWORD

Research Report Title:	Implementation of biological filtration for the treatment of cyanobacterial metabolites
Project Leaders:	Gayle Newcombe, Lionel Ho, Daniel Hoefel
Research Officers:	Emma Sawade, Benoit Lorain, Melanie Vache, Mike Dixon, Najwa Slyman, Martin Harris, Jennifer Dreyfus, Tim Tang
Research Nodes:	Australian Water Quality Centre
Project Partners:	Barwon Water, Coliban Water, Hunter Water Corporation, Melbourne Water, SA Water Corporation, United Water International and Veolia Water
Industry Team:	Joshua Feldman (Barwon Water), Dharma Dharmabalan (Coliban Water), Ray Stephenson (Coliban Water), Tim Giffin (Coliban Water), Raj Mahendrarajah (Coliban Water), Bruce Cole (Hunter Water Corporation), Mark Coleman (Hunter Water Australia), Benjamin Furness (Melbourne Water), Noel Miles (Melbourne Water), Nick Crosbie (Melbourne Water), Shane Hayden (Melbourne Water), Andrew van Prooyen (East Melbourne Laboratories), Keith Craig (Veolia Water), Colin Storey (Veolia Water) and Ali Gharaibeh (Veolia Water)
WaterRA Project No.	1021 – Implementation of biological filtration for the treatment of cyanobacterial metabolites

EXECUTIVE SUMMARY

The major water quality implication of cyanobacterial blooms is the production of secondary metabolites such as tastes and odours and a range of toxic compounds (cyanotoxins). In Australia many species of cyanobacteria proliferate in drinking water sources, in particular, species of *Dolichospermum*, *Microcystis* and *Cylindrospermopsis*. These cyanobacteria can produce the taste and odour compounds 2-methylisoborneol (MIB) and geosmin and cyanotoxins, specifically saxitoxins, microcystins and cylindrospermopsin (CYN).

Increasing frequency of cyanobacteria detection, coupled with expected global climate change, make it imperative that water authorities employ and optimise the multi-barrier approach for the mitigation of cyanobacteria and their metabolites. One barrier that has recently received widespread attention is the application of biological filtration for the removal of these metabolites.

Biological processes have been utilised for centuries for the treatment of drinking water. They are low technology processes that generally require little or no maintenance or running costs, and have the advantage of being a "natural" treatment, without the addition of chemicals or the possibility of harmful by-products. Biological removal of metabolites of cyanobacteria was studied extensively during CRC project 2405, "Biological Filtration Processes for the Removal of Algal Metabolites" (CRC project report no. 64). During that project, and an associated AwwaRF/CRC partnership project, "Towards the Development of Biosensors for the Detection of MIB and Geosmin", a number of outcomes were achieved that are of significant benefit to the water industry. While these two projects successfully achieved most of their stated objectives, the information was not sufficient to establish design criteria and operating guidelines for the implementation of biological filtration processes for the removal of cyanobacterial metabolites in operating water treatment plants. Although it has been shown that biological filtration is achievable at full scale, the effect appears to be site-specific, and we are yet to identify all the factors that will allow the implementation of this process at a range of sites under varying conditions. In particular, the time taken for a degrading biofilm to establish (the lag period) has been shown to be extremely variable, and this is a major consideration for the effective and confident application of biological filtration as a treatment barrier.

The objective of this project was to develop operational guidelines for the implementation of biological filtration for the removal of cyanobacterial metabolites at water treatment plants (WTPs).

The research approach was designed to address five hypotheses:

1. All filters have the potential to function as biological filters, given the right conditions
2. Media surface character and particle size will influence biodegradation
3. The existing potential at the full scale can be measured in the laboratory
4. Operations can be modified to allow biological filtration to occur
5. The lag period can be reduced by seeding with degrading bacteria

Summary of Outcomes

Identification of optimum operating conditions to enable effective biological filtration (addresses hypotheses 1 and 2)

Previous research shows that a range of filter media, of different composition and size, will support a healthy biofilm in the absence of a disinfectant residual. Various studies suggested that the topography and size of the media influence the formation of a biofilm, with smaller particle and rougher surface facilitating the development of a higher biomass.

In this study the aim was to compare different particle sizes of the same medium, and the same particle size of different media to determine the effect of filtration media particle size and surface characteristics on the implementation of biological filtration for the removal of metabolites. This effect had not been studied systematically previously. In laboratory-scale columns we investigated sand, filter coal and granular activated carbon (GAC), of two particle sizes each, and one other substrate that is promoted as a specialised filtration medium made from expanded clay particles (Filtralite). The expanded clay particles offered a novel surface chemistry, with pores of a size that should also be amenable to biofilm formation.

The laboratory columns were fed with settled water from the Happy Valley WTP (near Adelaide, South Australia) for 10 months. During this time the columns were backwashed weekly, and spiked periodically with cyanobacterial metabolites MIB, geosmin, saxitoxins, microcystin LR and LA (mLR and mLA) and cylindrospermopsin. Although all media developed a healthy and diverse biomass, as shown by scanning electron microscopy, the removals of the metabolites and natural organic matter (NOM) were at levels representative of physical processes, with no evidence of biological degradation. Under these conditions the GAC was effective for MIB and geosmin removal, with an average of 98 and 100% removal respectively over the 10 month study. In agreement with previous findings, the removal of the toxins was high at the time of commissioning (100% for microcystins and cylindrospermopsin and 85% for the saxitoxins, measured as saxitoxin equivalents (STXeq) by ELISA) then decreased over the 10 month trial. The reduction in adsorption for the toxins was in the order mLA>mLR>cylindrospermopsin>STXeq. At the completion of the trial the removal of mLA by both GACs was of the order of 20%; only slightly higher than the physical removals displayed by the non-porous media.

In previous pilot trials at Happy Valley biological removal was observed on sand after 10 months. After 10 months the current comprehensive trial was terminated with the conclusion that the different particle sizes and surfaces tested did not influence the development of a biofilm capable of degrading any of the metabolites over this period. While the removals of MIB and geosmin observed on the GAC filters at the conclusion of the trial may have been influenced by biological as well as physical processes, the removal of the toxins was through physical processes alone.

Two of the media were retained for investigation of the lag period (in this study defined as the length of time to achieve > 50% removal of metabolites) prior to the commencement of effective MIB and geosmin biological filtration.

Pilot plant investigations were undertaken at the Morgan WTP using sand from the full-scale filters. From previous column work it was assumed that the filters were functioning as biological filters and would remove both MIB and geosmin to below detection. This investigation was designed to determine the effect of hydraulic loading on the removal of the MIB and geosmin in a full-scale biological sand filter.

At filtration rates equivalent to normal full-scale operating conditions removals of both taste and odour (T&O) compounds were consistent with physical removal only, no additional removal attributable to biological processes was observed. This was unexpected as the plant had continued to remove geosmin effectively during recent challenges. However, during several episodes of high MIB levels entering the plant (up to 63 ng/L), only 40-60% MIB removal was observed at the full scale, some of which could be attributed to powdered activated carbon application. At lower filtration rates removals of geosmin increased with increasing contact time. These results were consistent with biological removal, although the observed removals were lower than expected from previous work. The differences observed between the full-scale removal of geosmin and at the pilot scale may be a result of the media. In the pilot columns only sand was used whereas the Morgan filters are sand/anthracite. It is possible that higher biological removals are experienced within the biofilm on the anthracite and this effect would not have been observed in this study.

Development of the "Biological Filtration Potential Test" (addresses hypotheses 1 and 3)
Effective biological removal of MIB and geosmin had been observed at Morgan WTP for over 14 years. The only previous interruption to this removal occurred during a period where monochloramine was present in the backwash water and the biofilm was damaged. Once the process was reversed the biofilm re-established within a few months and geosmin was again effectively removed. This effect was also seen at the full scale for MIB, although the removals were lower than for geosmin. Research over the past 10 years has shown conclusively that the filters at Morgan function as biological filters for MIB and geosmin, and it was hypothesised that other rapid filters could function in a similar manner under the appropriate conditions. However, at the full scale it is difficult to determine what processes in the plant are contributing to the removal of metabolites unless systematic investigative sampling is undertaken during a T&O or cyanotoxin event. A laboratory-scale procedure was developed to determine the extent of biological degradation occurring in existing filters in treatment plants managed by partner organisations, and the potential for a biofilm to develop that may be capable of degrading the metabolites. Five treatment plants were tested using the procedure, which included the metabolites MIB, geosmin, mLR and LA, STXeq and cylindrospermopsin. Only Morgan WTP sand

showed evidence of biological filtration and only for geosmin. The results also showed that at the time of sampling there were insufficient degrading bacteria in the influent to the filters at Morgan WTP to degrade the compounds in a batch experiment undertaken over a two week period. A similar batch experiment undertaken on the raw water to the plant showed the presence of MIB and geosmin degraders. It was suggested that process improvements implemented at the plant to reduce filtered water turbidity may have resulted in improved bacteria removal through optimised coagulation. At Grahamstown WTP (Hunter Water) the influent to the filters contained geosmin degrading bacteria; however the filter media did not display removal as a result of biodegradation. The samples from the other three plants showed no evidence of degrading bacteria in the influent water to the filters or in the filter media itself.

Investigation of potential impacts on the extent of the lag period for MIB and geosmin removal (addresses hypotheses 4 and 5)
Two of the filter media tested in the experiments undertaken at Happy Valley were used for further investigations. In addition, samples of anthracite and sand from the full-scale filters at Morgan WTP were tested in laboratory-scale columns. The aim of this study was to identify methods to enhance the removal of MIB and geosmin through biological filtration, and determine operational factors that could negatively impact the removals.

Three scenarios that may occur at the full scale were investigated:

Scenario 1 – Presence and absence of metabolites
It has been suggested in the literature that the rate of development of a biofilm capable of degrading metabolites (or the extent of the lag period) is influenced by the presence or absence of the metabolites themselves. In this project, samples of anthracite and sand from Morgan WTP filters were used for laboratory-scale column testing using Morgan settled water spiked with geosmin and MIB as the influent. The initial testing took place over 130 days. The results indicate that within 8 days both the sand and anthracite media removed greater than 50% of both T&O compounds. This indicates the lag period was short and prolonged exposure to metabolites can increase the removal. On day 43 the removal of both compounds was significantly impacted, especially in the anthracite media where removal decreased to physical removal levels. Continued running of each of the columns showed an increase in MIB removal from 15% to 74% for the anthracite media and 55% to 91% for the sand over approximately 90 days. Geosmin removals also increased, from 39% to 85% for anthracite and 75% to 90% for the sand. The cause for the large impact on the biofilm was unknown as the water quality parameters and operating conditions remained the same throughout the trial.

After 130 days, when the degrading biofilm had re-established, the columns were chlorinated to deactivate the biofilm, and the spiking was resumed at the previous levels. The observed lag period for geosmin was short, with greater than 50% removal achieved for both sand and anthracite within a week, while it took longer (22 days) for the removal of MIB to reach 50%. The results suggested that the biofilm on both media could recover from a chlorination event within 4 weeks, similar to the results seen at the full scale. After 36 days the filters were again de-activated using chlorine and the trial was re-commenced in the absence of the metabolites. After a further 22 days MIB and geosmin were added to the feed water to determine removals. The aim was to compare the lag period observed in the presence and absence of the metabolites after inactivation of the degrading biofilm. The removals of both MIB and geosmin at 22 days after biofilm deactivation were lower when the biofilm was re-established in the absence of the metabolites. For both compounds the recovery of the biofilm appeared to be more rapid on the sand.

Scenario 2 – Seeding with indigenous bacteria
Although artificially seeding laboratory filters with cultured metabolite-degrading bacteria has shown some promising results, further laboratory, pilot- and full-scale investigations have suggested that such advantages may be short-lived. As a result, testing was undertaken to determine whether artificially seeding with indigenous bacteria, with higher numbers than would normally be found in the influent water, would have a positive effect on the degradation of metabolites. In addition, as shown in the previous section, the presence of MIB and geosmin on a continuous basis may promote the increase in the numbers of degrading bacteria in the biofilm, or increase the degradation rate, and thus improve the degradation efficiency of biological filters.

Two of the media columns were chosen from the Happy Valley filter media study for this work (sand 6.5 and filter coal 0.8-1.0) with the aim to seed with indigenous bacteria directly using Happy Valley raw water and improve biological removal. Batch experiments showed the presence of both MIB and geosmin degraders in Happy Valley Reservoir water.

The columns were fed for approximately 3 months on Happy Valley raw water to allow biofilm development in the absence of metabolites, followed by an additional 12 months with MIB and geosmin intermittently spiked during this period. MIB and geosmin were then spiked continuously into the reservoir feed water at a target concentration of 100 ng/L, considered to be a concentration that is within the normal concentration range causing an operational T&O challenge, over a period of 50 days. Removals of both compounds was initially greater than 90% in both filter media, however the removal decreased in the sand column for MIB after a week, while the removal for filter coal remained above 90%. During the same period the removal of geosmin through the sand filter decreased to below 70%, then recovered within three weeks to the same removal as the filter coal. Similar to scenario 1, no cause for this decrease in biodegradation could be ascertained. The results showed that the indigenous degrading bacteria had established within the biofilm and effective biological filtration of MIB and geosmin occurred initially. In addition, after damage to the biofilm of an unknown cause, the biofilm recovered over a period of approximately one month. In this scenario the biofilm on the filter coal appeared to be less susceptible to damage than the sand biofilm as both were subject to the same conditions.

Scenario 3 – Presence and absence of degrading bacteria
The results obtained in the project to date and full-scale WTP data suggested that changing water quality, in particular the number and type of bacteria entering the filter, could have a significant impact on the potential for the filter to achieve effective biological filtration. In this section of work the two columns used in scenario 2 were changed from Happy Valley raw water to sterilised raw water. Therefore the physical water quality remained the same (nutrients, NOM pH etc) but there were no active bacteria present in the influent to replenish the biofilm. The columns were fed with the sterile raw water for four weeks and then spiked with MIB and geosmin for three weeks to determine the effect of the absence of degrading bacteria in the feed water.

The results showed 95% removal of geosmin during most of the spiking period, with a small decrease in removal at day 21. The initial removal of MIB was 71% for sand and 96% filter coal – also indicative of biological. In contrast to geosmin, the removal of MIB decreased markedly during the spiking period and on day 21 the effluent contained over nine times the concentration of MIB present in the influent. The results suggested that the absence of degrading bacteria in the influent for four weeks caused a change in bacterial population in the biofilm to a situation where producers of the compounds were established. After the spiking period the filters were fed with non-sterile Happy Valley Reservoir water for an additional four weeks in an effort to re-establish MIB degrading bacteria within the biofilm. During this period high concentrations of MIB were evident in the backwash water and the filter effluent of both media while geosmin concentrations remained at background levels. Both media were then chlorinated to deactivate all bacteria, and the trials were resumed on non-sterile Happy Valley water spiked with MIB and geosmin. The removals increased over time for both metabolites, with the lag period found to be 22 days for MIB on the sand filter. After 29 days the filters were chlorinated to deactivate the bacteria and the columns were fed with non-sterile Happy Valley Reservoir water in the absence of metabolites. As observed with the Morgan columns the time taken to reach >50% removal of T&O was longer in the absence of the metabolites. In this scenario both media showed similar trends although the percent removals on the filter coal were slightly higher.

Conclusions and recommendations

The establishment of a biofilm on filter media takes place within several weeks of commissioning. However, the establishment of a biofilm capable of degrading cyanobacterial metabolites to acceptable levels for the consumer will take considerably longer.

The long-term removal of cyanobacterial metabolites in biologically active filters is dependent on a range of complex and interrelated factors including:

- The number of degrading bacteria present in the influent water
- The ability of those bacteria to colonise filtration media
- A critical mass of degrading bacteria being retained in the media
- The maintenance of water quality conditions conducive to a healthy, degrading biofilm
- The frequent/regular presence of the metabolites in the influent to the filters

Some of these factors can be controlled at the treatment plant and some are dependent on the raw water quality. In addition, those factors that may be influenced by plant operations could conflict with other water quality objectives. For example, optimisation of coagulation for improved particle removal may result in reduced bacteria numbers in the influent to the filters, and a consequent decrease in the efficacy of biological filtration. In addition, results at the laboratory and full scale suggest that the efficacy of biological removal may be impacted rapidly by unknown factors, resulting in unforseen breakthrough into the treated water.

As a result, it isn't possible to develop specific guidelines for the confident application of biological filtration for the removal of cyanobacterial metabolites at all treatment plants. However general recommendations that may increase the probability of successful biological filtration are given below, and suggestions are made to aid in the determination of the potential for the development of the process at a particular plant.

Can your filters achieve successful biological filtration?
If you answer "yes" to all of the statements below there is the potential for the formation of a degrading biofilm on your filters:

- Our filtration medium has an average particle size < 1 mm
- There is no pre-chlorination or pre-filter chlorination at the WTP
- Metabolite(s) are regularly present in the raw water
- Metabolite(s) are regularly present at the inlet to the filters
- There are degrading organisms in the raw water
- Degrading organisms regularly reach the inlet to the filters

How can you improve the chances of biological filtration?

- Use smaller media particle size
- Cease pre-chlorination and/or pre-filtration chlorination if usually practiced
- Allow degrading bacteria to colonise filtration media – seed filters with indigenous bacteria if feasible
- If a degrading biofilm has established it will regenerate after severe damage comparatively quickly once suitable conditions for regrowth have been re-established

Granular activated carbon (GAC) is the recommended medium for the implementation of biological filtration for the removal of cyanobacterial metabolites. The adsorbent allows for the removal of metabolites during the often prolonged period prior to the onset of biodegradation. However, the same conditions apply as above – in the absence of these conditions adsorption of the cyanotoxins will decrease over time due mainly to the continuous adsorption of NOM. The GAC may be rendered ineffective by "pre-loading" of NOM over a period of 12 months or less.

CONTENTS

FIGURES

TABLES

1 INTRODUCTION

1.1 Introduction

Cyanobacteria have evolved to adapt to almost every environment on the planet, in particular to both marine and fresh water sources (Falconer 2005). They continue to adapt to our changing environmental conditions, and thrive in water sources impacted by development and climate change (Davis et al. 2009; Paerl et al. 2011; Paerl and Huisman 2008). A major trigger for the proliferation of cyanobacteria in water is the input of nutrients resulting from modern agricultural practices. In addition, warmer global temperatures, low flows in rivers and reduced water quality associated with droughts favour the growth of cyanobacteria (Paerl and Huisman 2008). As a result of all of these factors, issues associated with cyanobacteria in drinking water are increasing world-wide.

The major water quality implication of the proliferation of cyanobacterial blooms, for the drinking water industry, is the production of secondary metabolites, particularly those causing aesthetic issues (e.g. compounds which impart tastes and odours) and those which can severely impact human health (e.g. cyanobacterial toxins or cyanotoxins) (Carmichael 1992; Carmichael 1994; Chorus and Bartram 1999; Fleming et al. 2002). In Australia, many troublesome species of cyanobacteria proliferate in drinking water sources, in particular, species of *Dolichospermum*, *Microcystis* and *Cylindrospermopsis*. These cyanobacteria have the potential to produce the taste and odour compounds 2 methylisoborneol (MIB) and geosmin and/or the cyanotoxins – saxitoxins, microcystins and cylindrospermopsin.

Cyanotoxins, in particular microcystins, are susceptible to biodegradation (Christoffersen et al. 2002; Cousins et al. 1996; Jones and Orr 1994; Lam et al. 1995; Rapala et al. 1994; Saitou et al. 2002; Tsuji et al. 1996; Watanabe et al. 1992). A majority of studies have observed degradation of microcystins in natural waters (e.g. lakes and reservoirs).

Geosmin and MIB have also been shown to be readily biodegradable under a range of conditions (Elhadi et al. 2006; Ho et al. 2012a; Hoefel et al. 2009a; McDowall et al. 2007a; McDowall et al. 2007b; McDowall et al. 2009).

A growing number of microorganisms implicated in the biodegradation of microcystin and nodularin toxins have been isolated, and so far a majority appear to belong to the family *Sphingomonadaceae*. Within the genome of the first isolated microcystin-degrading bacterium, *Sphingomonas sp.* ACM-3962, Bourne et al. (1996; 2001) identified a gene cluster, mlrA-D, coding for enzymes responsible for the degradation of microcystin LR (mLR). The authors showed that the protein/enzyme MlrA is responsible for the hydrolytic cleavage of the cyclic structure of mLR. The resultant linear mLR molecule was then sequentially hydrolysed by peptidases MlrB and MlrC; MlrD, a putative transporter protein, may have allowed for active transport of microcystin and/or its degradation products into or out of the cell. Based on that work Saito et al. (2003) designed a PCR assay for detection of mlrA gene from *Sphingomonas sp.* ACM-3962. Using this assay, the authors reported gene homologues of mlrA in two microcystin-degrading bacteria, *Sphingomonas sp.* MD-1 and *Sphingomonas sp.* Y2, both of which were isolated from separate Japanese lakes.

Perhaps the best way to utilise microorganisms for the biodegradation of cyanotoxins is through biological filtration processes. Biological processes have been utilised for centuries for the treatment of drinking water. They are low technology processes that generally require little or no maintenance or running costs, and have the advantage of being a "natural" treatment, without the addition of chemicals or the possibility of harmful by-products. Of the biological filtration studies conducted on cyanotoxins, a majority have been via sand media. Lahti and Hiisvirta (1989) found promising results using a pilot-scale slow sand filter with up to 86% removal of microcystin (variant not determined) from *Microcystis aeruginosa*. Studies by Sherman et al. (1995), Grützmacher et al. (2002) and Ho et al. (2006) also reported efficient removal of microcystin in slow sand filters. Ho et al. (2006) provided evidence that the removal of microcystin through a sand filter was predominantly through biological processes and that removal also occurred under rapid sand filtration conditions. In addition, Ho et al. (2006) detected the mlrA gene within the biofilm of the sand filter using a PCR assay similar to that of Saito et al. (2003). This demonstrated the potential of the PCR assay to be used as a screening tool to assess the microcystin degrading capability of biologically active sand filters.

Biological removal of microcystins and anatoxin-a has also been reported in GAC filters (Carlile 1994; Fawell et al. 1993; Hart and Stott 1993; Ho and Newcombe 2007; Newcombe 2002; Newcombe et al. 2003; UKWIR 1996; Wang et al. 2007). GAC filters offer the advantage of two removal mechanisms, adsorption and biodegradation, and thus are an attractive treatment option for effective cyanotoxin removal. GAC filtration is often preceded with ozonation and this combination has been shown to be highly effective for microcystin removal (Craig et al. 1998; Newcombe 2002).

Studies on the potential for the saxitoxins to degrade in freshwater are limited. Jones and Negri (1997) discussed the persistence and degradation of saxitoxins in freshwaters and concluded that over a 90 day period, C-toxin concentration decreased exponentially (half-life of 21 days) while GTX2 and GTX3 also decreased with a half-life of 46 days. Ho et al. (2012b) investigated the biological degradation of the saxitoxins in two waters over two seasons and concluded the effect was negligible compared with the degradation of other metabolites. Biological filtration has been shown to be incapable of effectively removing the saxitoxins (Kayal et al. 2008). Kayal et al. (2008) assessed the fate of five saxitoxins variants (C1, C2, GTX2, GTX3 and STX) through biologically active filters containing sand and anthracite sourced from the filters beds of two water treatment plants. Decreases in the concentration of the less toxic variants coincided with increases in the concentrations of the more toxic variants through the filters containing anthracite while no changes in toxin concentrations were evident through parallel filters containing sand. The results strongly suggest that organisms within the biofilm of the anthracite filters possessed the ability to biotransform the saxitoxin variants, potentially to more toxic variants.

The operational conditions for biological filtration are important for the successful removal of cyanotoxins and T&O compounds. The type and concentration of bacteria, water quality, filter media, filter contact time and hydraulic loading all have a major impact on biological filtration processes. Perhaps the most difficult aspect of the application of this technology is the delay for biodegradation to commence. This is often referred to as the lag period or lag phase. Studies have reported lag periods of days to months before organic compounds are degraded. The lag period is a major hindrance for the application of biological filtration processes, particularly for the removal of contaminants that occur periodically such as cyanotoxins. In some cases lag periods can be reduced upon re-addition of the target compound. Senogles et al. (2002) and Smith et al. (Smith et al.) showed that cylindrospermopsin could be degraded by natural aquatic bacteria in a surface water. Senogles et al. (2002) found that after 30 days, cylindrospermopsin was degraded to below detection limit with a lag period of approximately 15 days. Upon re-addition of cylindrospermopsin to the sample they observed no lag period. Similarly, Rapala et al. (1994), Christoffersen et al. (2002), Holst et al. (2003) and Newcombe et al. (2003) have shown that for microcystin biofiltration, lag periods could be substantially reduced, in some circumstances eliminated, when bacteria were repeatedly exposed to the toxin.

- For further literature information on biological filtration for the removal of cyanobacterial metabolites refer to review article - Ho L, Sawade E, and Newcombe G (2012) Biological treatment options for cyanobacteria metabolite removal – A review, *Water Research*, 46 (5): 1536-1548.

1.2 Project background

Biological removal of metabolites of cyanobacteria such as the taste and odour (T&O) compounds MIB and geosmin, and cyanotoxins was studied extensively during CRC project 2405, "Biological Filtration Processes for the Removal of Algal Metabolites" (CRC project report no. 64). During that project, and an associated AwwaRF/CRC partnership project, "Towards the Development of Biosensors for the Detection of MIB and Geosmin" the following significant knowledge was gained:

General

- Biofilms will establish rapidly on any surface in contact with water
- Reaching a critical mass of degrading bacteria will result in degradation of metabolites
- Removal of the majority of the known cyanobacterial metabolites to below detection is achievable by sand filtration alone on a laboratory scale (e.g. Figure 1)
- Once an existing biofilm has been damaged, under appropriate conditions it will reform in a relatively short time (full scale and laboratory scale) (e.g. Figure 2)
- New media may take a substantial time to establish a biofilm capable of degrading the compounds (laboratory and pilot scale) (e.g. Figure 3)
- Media particle size (and therefore biofilm surface area) appears to have a significant effect
- A range of organisms are capable of degrading the compounds, using synergistic or individual degradation mechanisms
- Different bacteria appear to be responsible for the degradation of different metabolites (genera *Sphingomonad*)
- Under most conditions the onset of biological degradation, either in filters or batch experiments, displays a "lag" phase, during which there is either no degradation, or a gradual increase
- Seeding of degrading organisms may be beneficial in the enhancement and/or acceleration of metabolite removal

MIB and geosmin

- GAC is a very effective biological substrate and can function indefinitely if the biofilm remains robust (full scale and laboratory/pilot scale)
- Rapid sand filters can function in biological mode and remove MIB and geosmin to below detection, removing reliance on powdered activated carbon if the biofilm remains robust (full scale and laboratory/pilot scale) (e.g. Figure 4)

Cyanotoxins

- Microcystins – similar to MIB and geosmin, although no full-scale evidence, degradation appears to occur abruptly, but often after a lag phase that can be reduced by multiple spiking
- Cylindrospermopsin – can be degraded in batch experiments and on pilot slow sand filters, no evidence for rapid sand filters
- Saxitoxins – no evidence of removal due to biodegradation. Limited evidence of bio-transformation to more toxic forms
- Genetic techniques have been successfully applied to quantify microcystin degrading bacteria present on a biological filter

While these two projects successfully achieved most of their stated objectives, the information gathered was not sufficient to establish design criteria and operating guidelines for the implementation of biological filtration processes for the removal of cyanobacterial metabolites in operating water treatment plants.

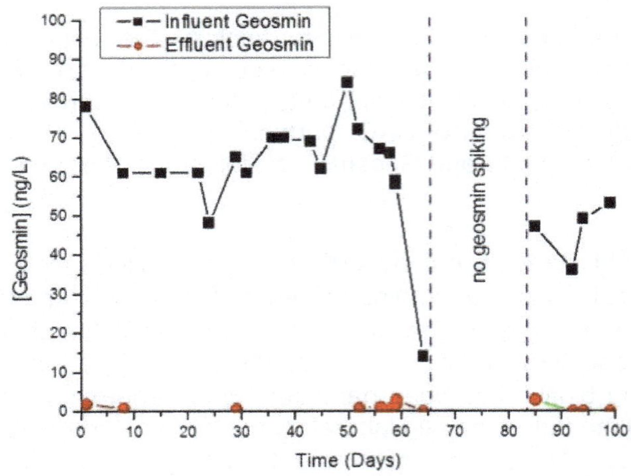

(a)

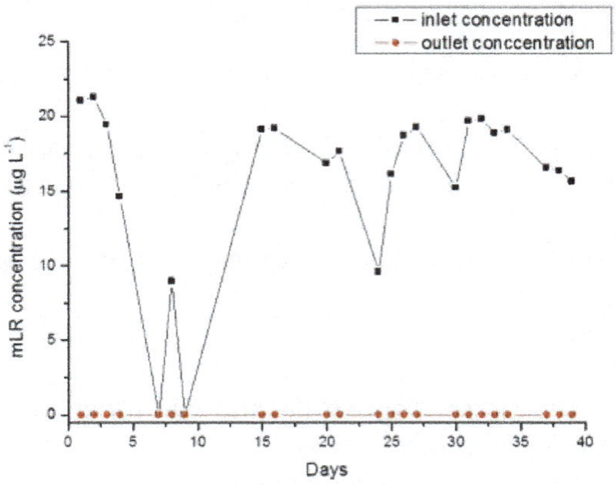

(b)

Figure 1: Removal of geosmin (a) and microcystin LR (b) to below detection in a laboratory-scale biologically active sand filter.

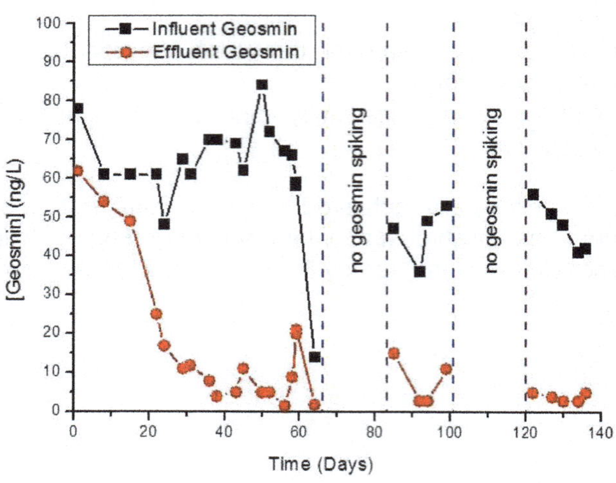

Figure 2: Evolution of biological removal with time after sterilisation of the biofilm. Recovery to 90% removal by day 40 after sterilisation.

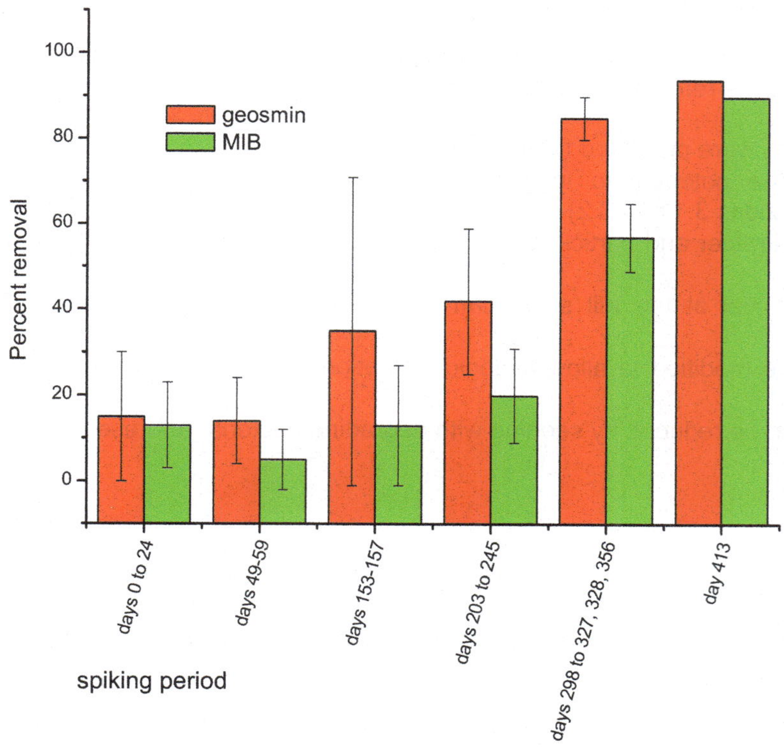

Figure 3: Removal of geosmin and MIB on a laboratory sand filter intermittently spiked with metabolites. Influent: Happy Valley settled water. Sand: effective size 0.55 mm. Total trial period: 413 days (McDowall 2008).

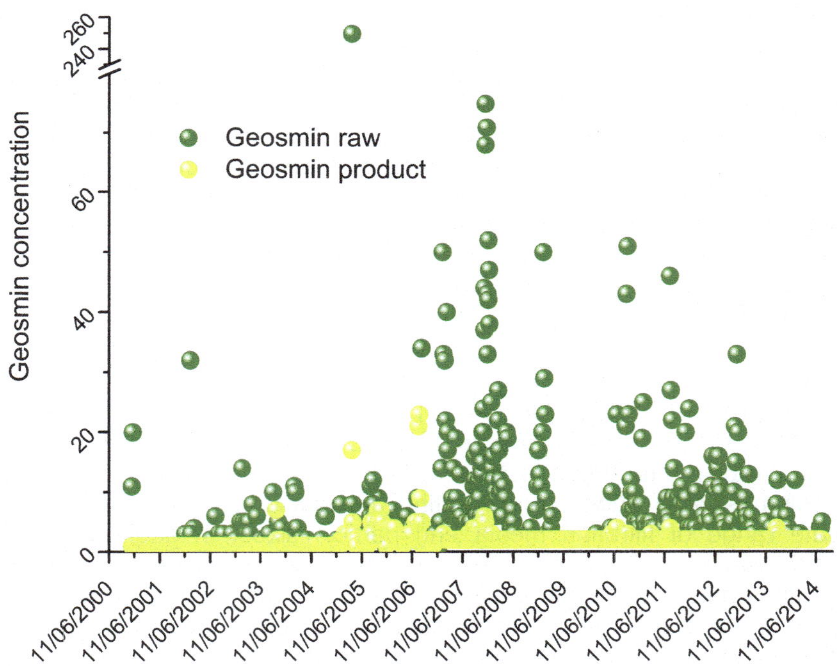

Figure 4: Removal of geosmin through biologically active rapid filters at Morgan Water Treatment Plant.

1.3 Aims and Objectives

The aim of project 1021 was to gather additional information to enable the implementation of biological filtration at the full scale for the removal of cyanobacterial metabolites.

Several project hypotheses were proposed to facilitate implementation:

- All filters have the potential to function as biological filters, given the right conditions (addressed in Sections 3.1 and 3.2)
- Media surface character and particle size will influence biodegradation (addressed in Section 3.1)
- The existing potential at the full scale can be measured in the laboratory (addressed in Section 3.2)
- Operations can be modified to allow biological filtration to occur (addressed in Sections 3.1 and 3.3)
- The lag period can be reduced by seeding with degrading bacteria (addressed in Section 3.3)

1.4 Approach

The project hypotheses were tested using the following approach, described below.

1.4.1 Identification of optimum operating conditions to enable effective biological filtration

This study was undertaken in a series of laboratory-scale column and batch tests. These tests have been shown previously to mimic accurately the performance of biofilters in the full scale (Ho 2004; McDowall et al. 2007a).

1.4.1.1 Identification of the most effective substrate for biofilm formation

From previous work, particle size, chemical composition and roughness (topography) of the media surface have been identified as important factors influencing biofilm growth and biological removal of contaminants (Elhadi et al. 2006; Hattori 1988; McDowall et al. 2007b; McDowall et al. 2009; Wang et al. 2007). In this work the aim was to compare different particle sizes of the same medium, and the same particle size of different media to determine the effect of filtration media particle size and surface characteristics on the implementation of biological filtration. This effect has not previously been studied systematically. Differences in physical removal and biological removal of cyanobacteria metabolites through laboratory-scale filter columns has been reported in the literature (Elhadi et al. 2004; Ho et al. 2007a; Ho et al. 2006) and a recent review of biological treatment options for cyanobacteria metabolites was conducted by the research team (Ho et al. 2012a). In laboratory-scale columns we investigated sand and anthracite (two particle sizes of each), GAC and one other substrate, Filtralite, promoted as a specialised filtration medium made from expanded clay particles. The expanded clay particles offered a novel surface chemistry, with pores of a size that should also be amenable to biofilm formation.

Ten laboratory-scale columns were set up and run at the Happy Valley Water Treatment Plant, fed from the settled water ducts. This water was chosen as it has been shown to contain T&O degrading bacteria, and its proximity to the Australian Water Quality Centre (AWQC) allowed easier access to the large volumes of water required for these tests. Initially six columns were commissioned. Five months into the trial two columns were added, representing an additional GAC and sand particle size to ensure an adequate range of filtration media and particle sizes. In addition, two columns were commissioned with the same media as existing columns to confirm the results of the original column work. Column conditions and filter media details are given in Table 1 and Table 2 respectively. The columns were intermittently spiked with cyanobacterial metabolites over a period of 10 months to monitor the evolution of biological removal on the various substrates.

Table 1: Laboratory-scale filter column conditions

Column Conditions	Parameters
Column internal diameter	2.5 cm
Filter medium bed height	15 cm
Empty Bed Contact Time (EBCT)	15 min
Flow rate	5 mL/min
Filtration rate	0.6 m/hr

Table 2: Details of filter media

Column ID	Material	Supplier, product	Particle size (mm)
A	Sand	Riversands, 6	1.5-1.7
B	Sand	Riversands, 6.5	0.85-0.95
C	Bituminous coal	Riversands, filter coal	0.8-1.0
D	Bituminous coal	Riversands, filter coal	1.2-1.4
E	Expanded clay	Filtralite HC	0.8-1.6
F	GAC*	Acticarb, GAC1000N	0.5-0.7
G	Sand	Riversands,7	0.6-0.7
H	GAC*	Calgon, F300	0.8-1.0
B2	Sand (duplicate)	Riversands, 6.5	0.85-0.95
F2	GAC* (duplicate)	Acticarb, GAC1000N	0.5-0.7

Granular activated carbon

1.4.1.2 Determination of the effect of metabolite concentration on biological degradation

Several studies have reported increased rates of metabolite degradation with higher initial concentrations (Hoefel et al. 2009a; Saitou et al. 2002; Smith et al. 2008). During the spiking periods undertaken at the laboratory scale at Happy Valley WTP concentrations of metabolites, MIB, geosmin and cyanotoxins, were varied within the range that might be expected to challenge WTP operations under Australian conditions.

1.4.1.3 Effect of temperature on biodegradation of saxitoxins and cylindrospermopsin

Information in the literature, and previous work at the AWQC, show that the temperature range for effective biological degradation of MIB, geosmin and microcystin is approximately between 11 and 28°C, indicating that biological removal is feasible in most Australian conditions (Christoffersen et al. 2002; Elhadi et al. 2006; Ho et al. 2007b; Hoefel et al. 2009a). Previous studies at the AWQC have indicated that saxitoxins are recalcitrant to degradation, and some of the variants may undergo a biological transformation to more toxic variants, thus resulting in an increased overall toxicity of the treated water (Kayal et al. 2008) No information is available at present regarding the effect of temperature on the biodegradation of saxitoxins. In this study, batch biodegradation experiments in two waters, Swan Reach settled water (Murray River water location) and Happy Valley settled water, were undertaken at temperatures of 5°C, 15°C, 20°C and 30°C to determine if degradation or biological transformation of these compounds was a function of temperature.

From the limited data available on the effect of temperature on the biodegradation of cylindrospermopsin (CYN), the maximum degradation rate appears to be around 25°C, with a much slower rate at 20°C (Smith et al. 2008). CYN has been shown to be relatively recalcitrant to biodegradation, with slower degradation rates than microcystin (Ho et al. 2007b; Smith et al. 2008). In those CYN degradation studies, various waters were tested and degradation occurred only in water sampled from North Pine Dam in Brisbane, whereas none occurred in three South Australian source waters over a period of 250 days. The authors concluded that, as the SA waters had not experienced a previous *Cylindrospermopsis* bloom, no indigenous bacteria had developed the ability to degrade the toxin. Since that time the Murray River in South Australia has experienced several blooms of this cyanobacteria, although the limited monitoring did not identify the production of CYN. In this study batch biodegradation studies were undertaken in two waters, Myponga floated water and Swan Reach settled water (Murray River water location), using the indigenous bacteria present in the water samples at temperatures of 5°C, 15°C, 20°C and 30°C. Additional, batch experiments were also completed in Swan Reach raw water at 15°C and 20°C.

1.4.1.4 Determination of the effect of hydraulic loading on biological degradation of MIB and geosmin

Initially it was proposed that two effects of flow, measured by empty bed contact time (EBCT), were to be assessed:

- Effect of decreasing EBCT on biological removal through existing biofilms
- Effect of low EBCT on formation of biofilm within the filters

From the laboratory-scale column study undertaken at Happy Valley (identification of most effective substrate, as described in section 1.4.1.1) it was observed that, at an EBCT of 15 minutes a biomass capable of degrading the metabolites was not developed over a period of 10 months. These results suggest that lowering the EBCT may increase the time to form a biofilm on the media, hence taking longer than 10 months. Due to this potentially significant biomass formation time on new media, an alternative filter medium, sourced from the Morgan Water Treatment Plant (WTP) was used.

Morgan WTP services a large portion of regional South Australia. The plant operates with conventional treatment comprised of coagulation, flocculation, sedimentation, filtration and monochloramine disinfection. Morgan WTP receives water from the Murray River, which has frequent cyanobacteria blooms and regular mid to high levels of geosmin (Figure 4). More recently MIB has also been an issue in the inlet to this plant (mean value since 2011 is 10 ng/L with a maximum of 63 ng/L). Despite the occasionally high concentrations of geosmin in the raw water, the Morgan WTP has historically produced water with metabolite concentration close to, or below, detection (see Figure 4). Morgan WTP utilises dual-media anthracite-sand rapid gravity filters, with no pre-chlorination. Due to the long period of time in operation and the absence of chlorine the filters contain a diverse range of organisms and hence Morgan filter sand was chosen as the alternative media for this study.

This investigation was undertaken at the pilot scale at Morgan Water Treatment Plant (WTP). Following commissioning the pilot columns were run over a period of a few months prior to spiking with T&O compounds. The feed water (settled water from the WTP) was spiked continuously with T&O compounds during the trials, which lasted a week for each EBCT. As the filtrate of the columns was returned to the sludge treatment facility from which the supernatant is returned to the plant, only the T&O compounds were investigated.

1.4.2 Development of the "Biological Filtration Potential Test"

The research team has vast experience in laboratory-scale biological degradation tests. Previous studies in our laboratory showed that small-scale column tests can be utilised effectively to simulate the processes of adsorption and/or biological degradation in full-scale and pilot filters (McDowall 2008). In this section of work we have identified standard experimental testing conditions that offer the best indication of the potential for biodegradation within full-scale filters and developed and applied a testing protocol – the Biological Filtration Potential (BFP) test, Appendix I.

Five samples of filtration media and filter inlet water, sourced from project partners, were tested using the BFP test. The sites were chosen as they experience periodic episodes of MIB and/or geosmin in the source waters and chlorination is not practiced in the treatment process prior to disinfection.

The testing was undertaken in four parts, laboratory-scale filtration, batch testing, microbiological analysis and particle size observation. A brief description of the aims of these tests is given below.

Laboratory-scale filtration
Two small columns were run in parallel, spiked with a range of metabolites. One column holds the filter medium as sampled, the other holds a sterilised sample. The test was undertaken over a period of 12 days during which the inlet and outlet concentrations are determined. From this test the removal to be expected from physical processes (such as volatilisation, adsorption) is determined from the removal by the sterile sample, and any biological removal is calculated from the difference in removal between the sterile and non-sterile samples.

Batch experiments
Batch degradation experiments were undertaken on the influent water to the full-scale filters. Two samples of water were spiked with a range of metabolites; one was sterilised and the other was used as sampled. The concentration of the metabolites was determined over a period of 15 days and any difference in the concentration between the sterile and non-sterile samples is attributed to biological degradation. This test investigates the presence of potential degraders in the influent water to the filters.

Microbiological analysis
Two genetic approaches were employed for the microbiological analysis. DNA from the filter media was extracted using a commercially available kit (PowerSoil DNA Isolation Kit, MO BIO Laboratories, USA). Each DNA extraction was then screened for the abundance of the 16S rRNA gene fragment by using the primer set 27F-1492R for qPCR amplification (Hoefel et al. 2005). Results of this test provide information on the total population of micro-organisms present in the filter media. The extracted DNA was also screened for a specific gene involved in microcystin degradation, mlrA, using a previously published qPCR assay (Hoefel et al. 2009b). The presence of this gene indicates the presence of potential microcystin-degrading bacteria in the filter biofilm.

Particle size observation
For this test the particles were photographed beside a 1 mm graduated scale to estimate and compare particle sizes. Particle size has been shown to have an effect on biomass, and possibly the biological filtration potential. Smaller particles have a larger surface area, and therefore can potentially sustain a higher biomass than the same volume of medium of a larger size.

1.4.3 Investigation of potential impacts on the extent of the lag period for MIB and geosmin removal

The delay that is sometimes observed prior to the onset of degradation of several metabolites, even in the presence of an established biofilm, is often called the "lag phase" although this can be somewhat misleading. Extensive laboratory-scale degradation studies have shown that the onset of biodegradation of cyanobacterial metabolites can occur in several ways. Examples are shown in Figure 5 (a) – (d), where the y-axis represents percent removal of the metabolites, and the x-axis represents time. For example, Figure 5 (a) shows the immediate onset of 100% removal of the metabolite with no lag phase (ideal). In Figure 5 (b) there is a definite lag period prior to rapid degradation. However in Figure 5 (c) and (d), the period prior to the onset of biodegradation is less well defined. The definition of lag period in terms of microbiological systems is the period of low or no cell number increase preceding the exponential growth phase. For example when a bacterial inoculum is placed in a different medium, the lag period is associated with a physiological adaptation by the cells to the new environment prior to their resumption of division (Figure 6). In the case of biological degradation of metabolites the delay probably involves adaption, or acclimation, of the cells to the new food source, an increase in the number of specific bacteria capable of degrading the metabolite to a point where the removal is measurable, or a combination of both.

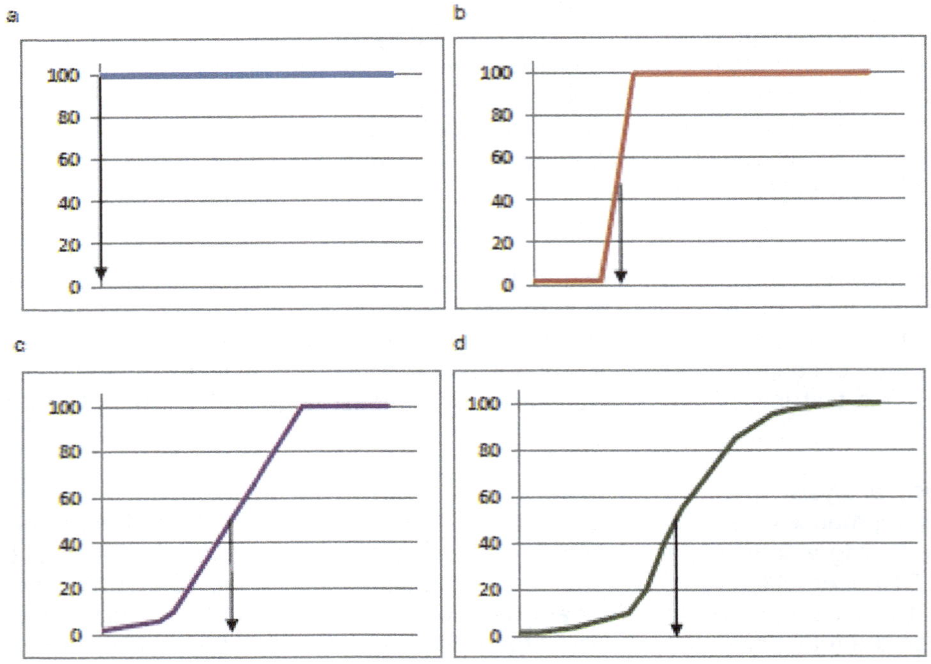

Figure 5: Examples of onset of biological degradation in a sand filter, percent removal versus time.

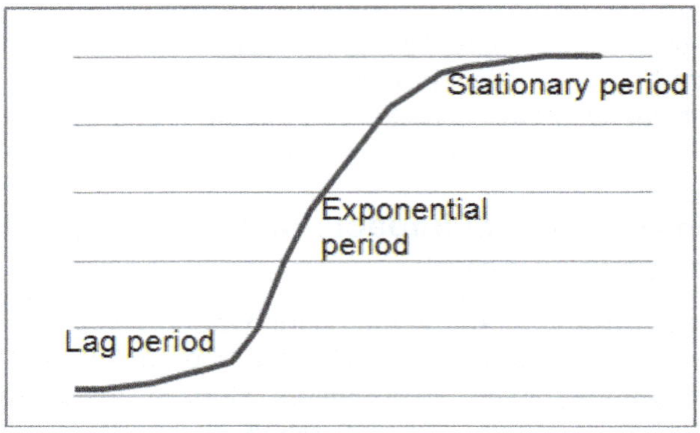

Figure 6: Bacterial growth curve, cell number versus time.

For the confident application of biological filtration to remove cyanobacterial metabolites, the preference is for no delay, as shown in Figure 5 (a). However, in most cases the behaviours shown in Figure 5 (b) – (d) will be apparent. For the practical, cost-effective application of biological filtration (eg as a replacement for PAC application) a removal of at least 50% of the cyanobacterial metabolites would be expected. Therefore, for the purposes of this project, the lag period in biological filtration was defined as the time for the column to reach 50% removal after the introduction of the metabolites into the system i.e. the times indicated by the arrows in Figure 5. This removal has also been shown to be above that reported for physical removal of metabolites in laboratory and pilot-scale columns (Elhadi et al. 2004; Ho et al. 2007a; Ho et al. 2006).

Previous work by the research team resulted in the isolation of a bacterium capable of degrading microcystin analogues mLR and mLA (Ho et al. 2007b) and geosmin (Hoefel et al. 2009a). It was envisaged that such bacteria could be seeded into filters to enhance the removals of cyanobacterial metabolites and potentially eliminate any lag period before efficient removals are achieved. The research team has already demonstrated the feasibility of such an approach where the seeding of a geosmin degrading bacterial consortium onto laboratory-scale sand filters resulted in the enhanced biofiltration of geosmin (McDowall et al. 2009).

Although artificially seeding laboratory filters with cultured metabolite-degrading bacteria has shown some promising results, further laboratory, pilot and full-scale investigations have suggested that such advantages may be short-lived. The long-term removal of cyanobacterial metabolites in biologically active filters is expected to be dependent on:

- The number of degrading bacteria present in the influent water
- The ability of those bacteria to colonise filtration media,
- The number of degrading bacteria retained in the biofilm
- A critical mass of degrading bacteria being retained in the media and replenishment after expected losses during the normal backwash operations
- The maintenance of water quality conditions conducive to a healthy, degrading biofilm

The "once-off" artificial introduction of non-indigenous degrading bacteria into a full-scale filter would not satisfy these necessary conditions. As a result, testing was undertaken to determine whether artificially seeding with indigenous bacteria, with higher numbers than would normally be found in the influent water, would have a positive effect on the degradation of metabolites. In addition, it was proposed that the presence of MIB and geosmin on a continuous basis may promote the increase in the numbers of degrading bacteria in the biofilm and thus improve the degradation efficiency of biological filters.

Three scenarios were investigated at the laboratory scale to determine if seeding with indigenous bacteria would increase biological removal of taste and odour compounds, MIB and geosmin and whether the presence of these compounds increases the efficiency -

i) Presence and absence of metabolites
ii) Seeding with indigenous bacteria
iii) Presence and absence of degrading bacteria

The objective was to assess some potential effects of changes in water quality on the implementation of biological filtration at the full scale.

1.4.4 Research Flow diagram

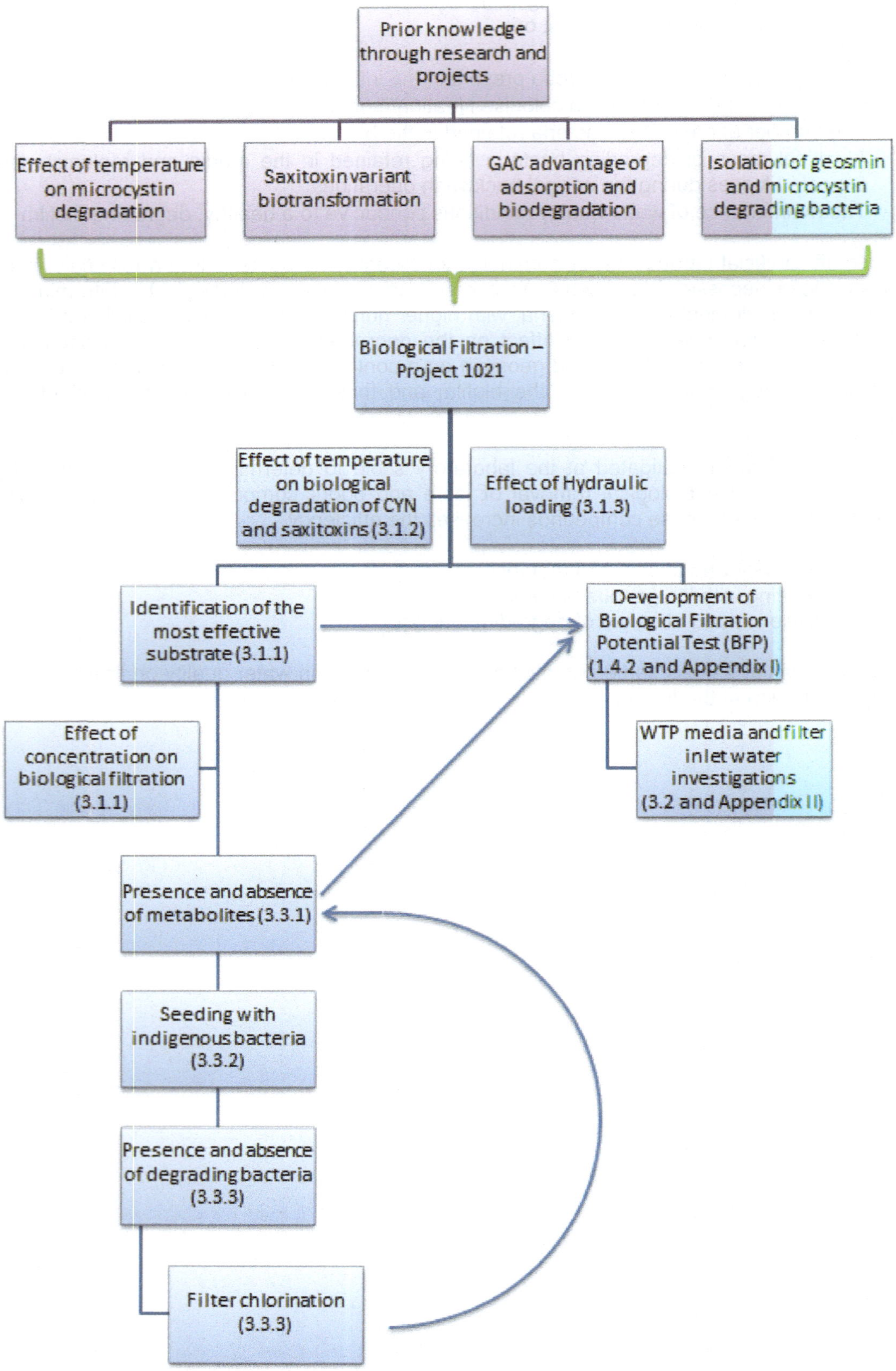

Figure 7: Flow diagram of investigation linkages within project.

2 MATERIALS AND METHODS

2.1 Cyanobacterial metabolite materials

Geosmin and 2-methylisoborneol (MIB) were purchased from a commercial supplier (Novachem Pty Ltd, Australia) and dissolved in ultrapure water (Millipore Pty Ltd, USA) to prepare a stock solution. Aliquots from the stock solution were then dosed into waters at the required target concentrations.

Cylindrospermopsin (CYN) was purchased from a commercial supplier (Sapphire Bioscience Pty Ltd, Australia) and dissolved in methanol and ultrapure water to prepare a stock solution. Aliquots from this stock solution were then dosed into sample waters at the desired concentrations.

Microcystin-LR (mLR) and microcystin-LA (mLA) was extracted from a natural *Microcystis aeruginosa* bloom, which occurred in a lake in Gippsland, Victoria, Australia, and was used for dosing during experiments. The extraction procedure involved freeze-thawing the bloom material in water and methanol, followed by preparative reverse phase flash chromatography and preparative high performance liquid chromatography (HPLC).

Five saxitoxin variants were used in this study: C1, C2, GTX2, GTX3 and STX. These were extracted from a natural bloom of *Dolichospermum circinalis* that occurred in Myponga Reservoir, South Australia. The isolation procedure involved freeze-thawing the bloom material of *D. circinalis* in 75% methanol. The toxin concentrate was purified and passed through ultrafiltration membranes to remove high molecular weight organics. The final stock solution (containing the five variants) was used in the experiments.

2.2 Analyses

2.2.1 Water quality analysis

2.2.1.1 Dissolved Organic Carbon (DOC) and UV absorbance

Samples for DOC and UV absorbance were filtered through 0.45µm pre-rinsed (ultrapure water) membranes. DOC was measured using a Seivers 900 Total Organic Carbon Analyser (GE Analytical Instruments, USA). UV absorbance at 254nm was measured through a 1 cm quartz cell using an Evolution 60 Spectrophotometer (Thermo Scientific, USA).

2.2.1.2 MIB and geosmin

MIB and geosmin samples were concentrated using a solid phase microextraction (SPME) syringe fibre (Supelco, Australia) and analysed on a Hewlett Packard 5890 Series II Gas Chromatograph with Hewlett Packard 5971 Series Mass Selective Detector (Agilent Technologies, Australia) against quantified labelled internal standards (Ultrafine Chemicals, UK) as described by Graham and Hayes (1998).

2.2.1.3 Cyanotoxins

Prior to high performance liquid chromatographic (HPLC) analysis, cylindrospermopsin was concentrated by solid phase extraction using methods described previously by Metcalf et al. (2002). An Agilent 1100 series HPLC system consisting of a quaternary pump (G1311A), degasser (G1379A), autosampler (G1313A), column compartment (G1316A) and photodiode array detector (G1315B) driven by ChemStation software (Agilent, California USA) was used for CYN analysis.

Microcystins were extracted from water by C18 solid phase extraction (Waters, 500 mg C18 cartridge) according to the method of Nicholson et al. (1994). The samples were analysed by HPLC with PDA detection at 240 nm.

Saxitoxin toxicity equivalents were measured using enzyme-linked immunosorbent assay (ELISA) purchased from a commercial supplier (Abraxis LLC, USA). Procedure was provided by the manufacturer with the ELISA kits.

For the temperature study all cylindrospermopsin and saxitoxin samples were analysed using commercially available ELISA kits (Abraxis LLC, USA), due to the volume of water required for analysis by HPLC. Prior to ELISA analysis the samples were diluted in a solution provided by the supplier to bring the concentrations within the working range of the assay (0.05-2.0 µg/L and 0.02-0.40 µg/L for CYN and saxitoxin, respectively).

2.2.2 Biological activity

2.2.2.1 Adenosine triphosphate

Adenosine triphosphate (ATP) was measured using a method outlined by McDowall (2008). Samples of filter media, from the laboratory-scale filter columns in the substrate investigation, were taken from the top of each column prior to and immediately after the spiking events.

2.2.2.2 Scanning electron microscopy

Samples of the filter media were taken after the 6 month event from the laboratory-scale filter columns in the substrate investigation. They were fixed for 1 hour using EM fixative (4% paraformaldehyde/ 1.25% glutaraldehyde in PBS, + 4% sucrose, pH7.2), washed in buffer (PBS, + 4% sucrose), post-fixed with 2% osmium tetroxide (OsO_4) in water for 30 minutes and then dehydrated with increasing ethanol solutions (70%, 90%, 100%). Subsequently the samples underwent critical point drying and were mounted on a stub and coated with carbon ready for analysis with a Field Emission Scanning Electron Microscope (FESEM). Samples of the virgin filter media were also mounted and coated ready for analysis. Photographs were obtained using a Philips XL30 Field Emission Scanning Electron Microscope.

2.2.2.3 mlrA gene and 16S rRNA abundance

Samples of the filter media that were analysed for mlrA gene and 16S rRNA abundance were processed as per procedures found in Hoefel et al. (2009b) and Hoefel et al. (2005), respectively.

2.3 Laboratory-scale filter column testing

2.3.1 Identification of the most effective substrate for biofiltration

Laboratory-scale tests were undertaken at Happy Valley WTP using conditions shown in Table 1 and the filter media listed in Table 2. Spiking trials were conducted at 0, 3, 6 and 10 months after commissioning to determine the removal of cyanobacterial metabolites. Two inlet concentrations of metabolites were used for the second, third and fourth trials. During the 1-2 week spiking trials, metabolites were dosed in continuously and samples were collected from the influent and filtrate daily for analysis. Percent removals were calculated, as were mean removals for each inlet concentration, and errors associated with one standard deviation. Table 3 gives the target concentration for the spiking trials. Figure 8 shows schematic of the column setup for the trials, Figure 9 shows a photograph of the 10 columns in the laboratory at Happy Valley WTP.

Table 3: Concentrations used for spiking periods, T = 0, 3, 6 and 10# months, for cyanotoxins and taste and odour compounds

Metabolites	**Time 0 months** (high concentration)	**Time 3, 6 and 10# months**
MIB	50 ng/L	Week 1 (low concentration): 15 ng/L Week 2 (high concentration): 50 ng/L
Geosmin	50 ng/L	Week 1 (low concentration): 15 ng/L Week 2 (high concentration): 50 ng/L
STXeq	15 µg/L	Week 1 (low concentration): 9 µg/L Week 2 (high concentration): 15 µg/L
Cylindrospermopsin	7 µg/L	Week 1 (low concentration): 4 µg/L Week 2 (high concentration): 7 µg/L
Microcystin (mLA and mLR mixture)	7 µg/L	Week 1 (low concentration): 4 µg/L Week 2 (high concentration): 7 µg/L

Completed for 6 columns only

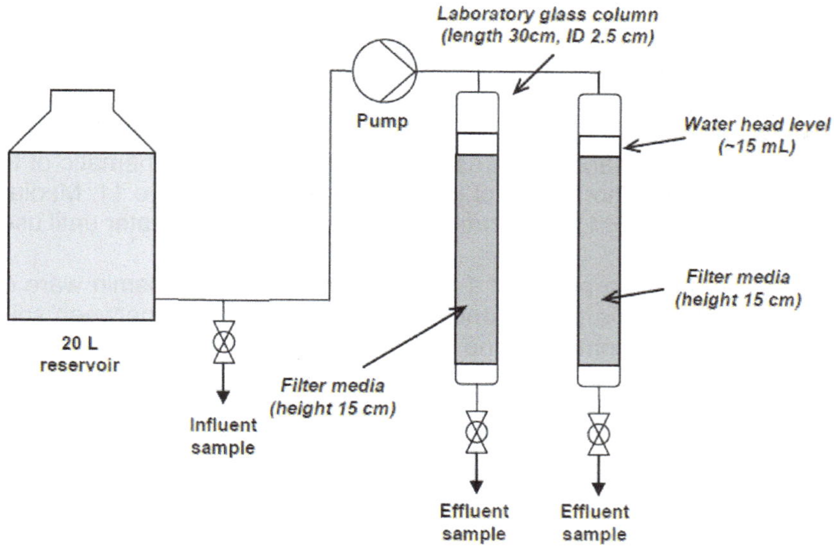

Figure 8: Laboratory column trial setup.

Figure 9: Laboratory columns, Happy Valley WTP.

2.3.2 Investigation of potential impacts on the extent of the lag period

Laboratory-scale tests were also utilised to investigate the source of lag period using conditions shown in Table 1. MIB and geosmin was spiked into feed water at a target concentration of 100 ng/L. During the spiking trials, MIB and geosmin were dosed into the feed water continuously and samples were collected from the influent and filtrate twice per week for analysis, with the exception of the work in the absence of metabolites. In the absence of metabolites, MIB and geosmin were only spiked into the feed water on day 22 and samples collected after a minimum period of four bed volumes.

2.4 Hydraulic loading investigation using pilot-scale filter column testing

Two columns (96mm diameter, 502mm bed height, bed volume 3.6L) containing Morgan filter sand were commissioned at Morgan WTP to assess the effect of EBCT on the removal of MIB and geosmin. Settled water (prior to full-scale filtration) was used as the feed water for both columns and filtrate water (prior to chloramination) was used for backwashing. A schematic of the pilot-scale filters is shown in Figure 10 while a photograph of the setup is shown in Figure 11. Media was collected from filter 4 at Morgan WTP (25/8/2011) and stored in buckets with settled water until use.

Four EBCT were studied over a period of 1 week where MIB and geosmin were continuously spiked into the influent of the filters at a target concentration of 100 ng/L. In between spiking trials, columns were fed with settled water containing no spiked contaminants. Samples were collected daily in duplicate over the spiking period from the influent and effluent of each column as well as from the settled water prior to spiking and 0.45 µm filtered for analysis by gas chromatography/mass spectrometry. Table 4 lists the conditions of the pilot-scale filters for the individual EBCTs to be assessed and the corresponding filtration rates (m/hr).

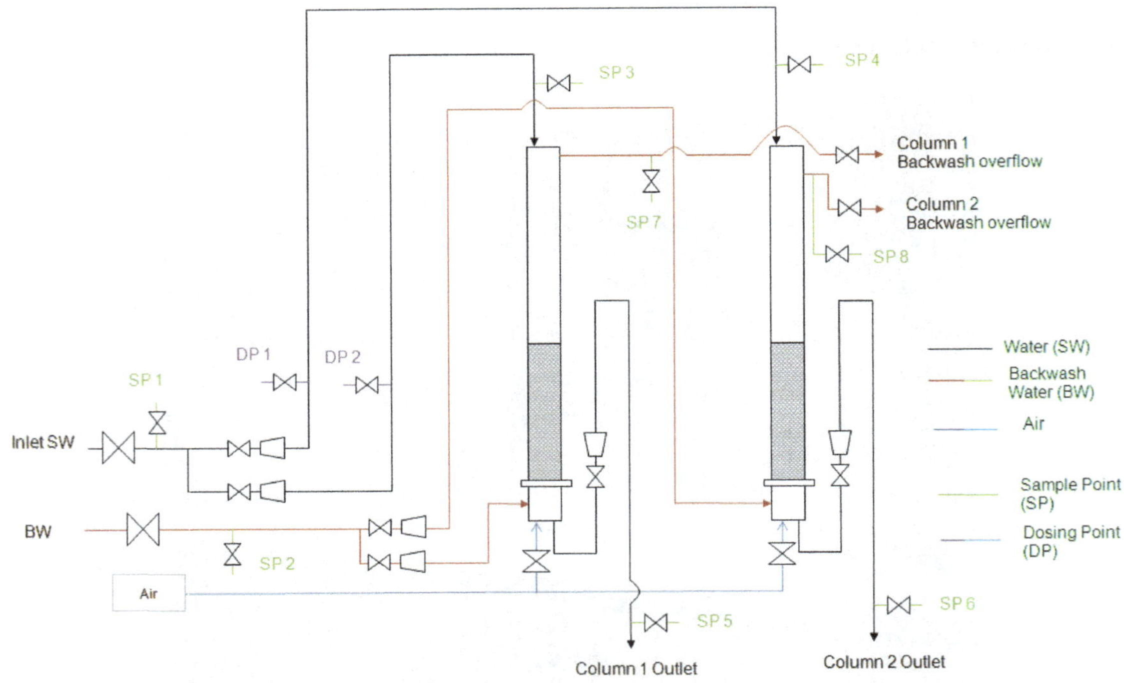

Figure 10: Schematic of the pilot-scale sand filters located at Morgan WTP.

Figure 11: Photograph of pilot-scale sand filter at Morgan WTP.

Table 4: Pilot-scale filter conditions

EBCT (minutes)	5	8	10	15
Flow rate (L/hr)	43.6	27.3	21.8	14.5
Flow rate (L/min)	0.73	0.45	0.36	0.24
Filtration rate (m/hr)	6	3.8	3	2

2.5 Effect of temperature on the biodegradation of saxitoxins and cylindrospermopsin

These experiments were undertaken in three waters sources, two for each metabolite. Each water was sampled at a full-scale plant at the inlet to the rapid sand filters. Myponga Reservoir water is coagulated with alum then floated prior to filtration, while Happy Valley Reservoir water, once coagulated with alum, is settled in sedimentation bays prior to filtration. The Murray River is the source for the Swan Reach WTP where the raw water is coagulated with alum, clarified in tube settlers then filtered.

Two reactors (2 x 1 L Schott bottles) were set up at each temperature for each water: one control (sterilised water sample – autoclaved 15 min at 121°C) and one batch (as is water sample). The reactors were placed in the appropriate incubator/fridge to equilibrate the water to temperature (24 hours) prior to spiking with the metabolite of interest. Each reactor was spiked with 10 µg/L of saxitoxins (volume of each reactor 1 L) and CYN (volume of each reactor 0.5 L).

Prior to collection of samples the reactors were inverted to simulate natural mixing. A volume of 1-2 mL was collected and 0.22 µm syringe filtered for analysis by enzyme linked immunosorbent assay (ELISA).

2.6 Biological Filtration Potential Test

Refer to Appendix I for Biological Filtration Potential (BFP) Test Procedure.

3 RESULTS AND DISCUSSION

3.1 Identification of optimum operating conditions to enable effective biological filtration

3.1.1 Identification of the most effective substrate for biofiltration

As discussed in the Introduction and the Materials and Methods sections a range of substrates were compared at the laboratory scale to determine the most effective substrate for biological filtration. The columns were monitored on an ongoing basis for natural organic carbon (NOM) removal, and spiked intermittently to determine metabolite removals.

3.1.1.1 Dissolved organic carbon and UV absorbance removal

The removals of dissolved organic carbon (DOC) and UV absorbance at 254 nm (UV) throughout the trial are shown in Figures 12 and 13. No removal was observed through the sand, filter coal or Filtralite filters, suggesting that if any physical or biological removal occurred the level was below the limit of detection of the analytical methods. The GAC filter was initially effective for the removal of these parameters, with the removal decreasing with time. This can be seen more clearly in Figure 14, which shows the percent removal of UV absorbing components over time. After approximately 100 days a semi-steady state removal of approximately 20% is observed. This is possibly due in part to some biological removal, but can be mainly attributed to physical removal by the highly porous GAC. These results are in agreement with previous laboratory and pilot-scale studies (Newcombe 2002).

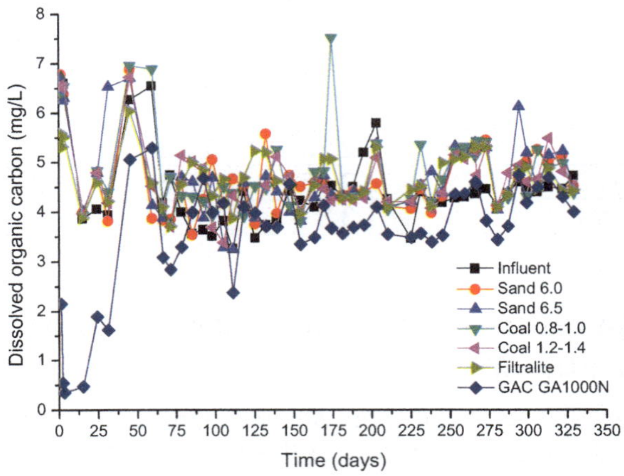

(a)

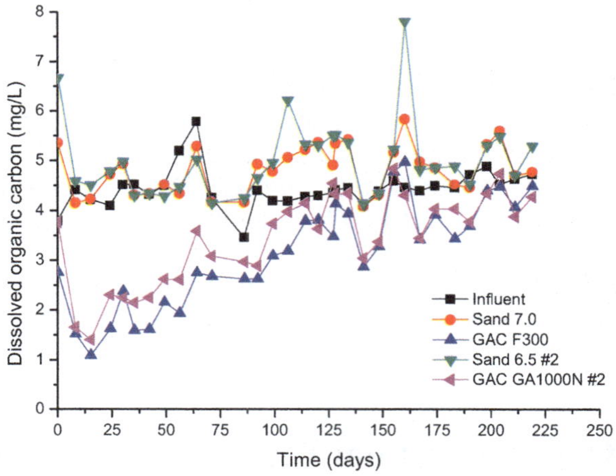

(b)

Figure 12: DOC concentration of influent and filtrate: initial 6 columns (a) and additional 4 columns (b).

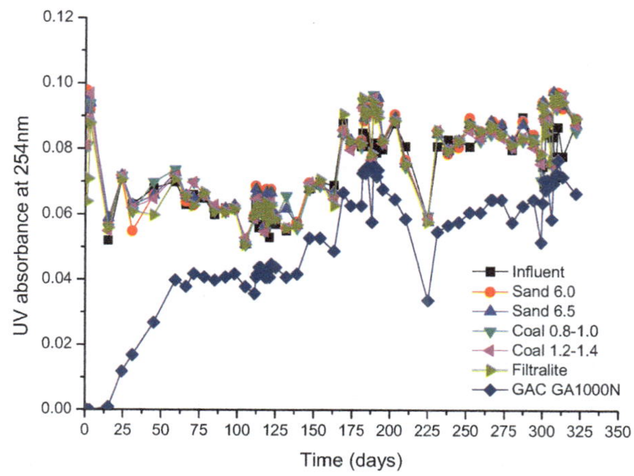

(a)

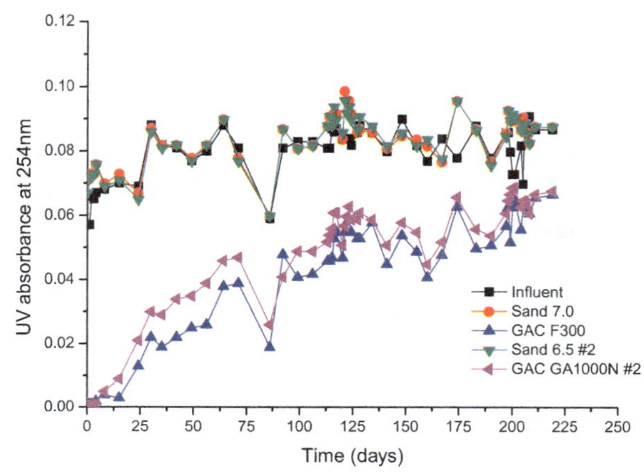

(b)

Figure 13: UV absorbance before and after filtration: initial 6 columns (a) and additional 4 columns (b).

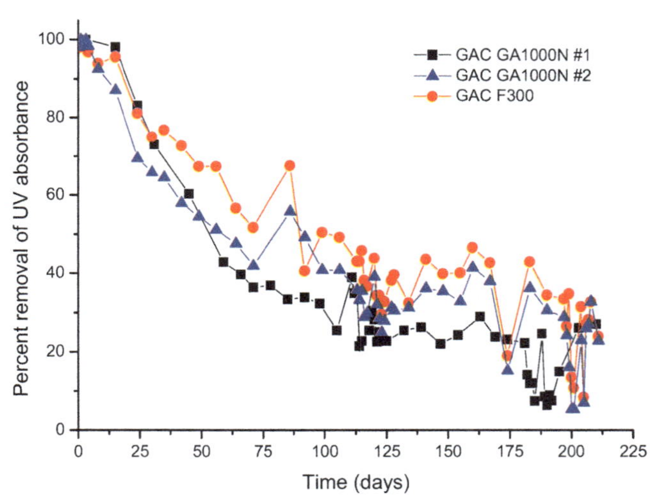

Figure 14: Percent removal of UV absorbance by GAC filter media.

3.1.1.2 Metabolite removal

For the initial six columns, four spiking trials were conducted over a 10 month period, at 0, 3, 6 and 10 months after commissioning, to determine the removal of cyanobacterial metabolites. Two inlet concentrations of metabolites were used for the second, third and fourth trials (refer to Table 3 for target concentrations). The average percent removals and standard deviations (sd) were calculated for each inlet concentration for each spiking period.

3.1.1.2.1 MIB and geosmin

The data for MIB and geosmin removals through the initial six columns is shown in Figure 15 and Figure 17 respectively.

An application of ANOVA (analysis of variance) to the sand, filter coal and Filtralite media data showed that there was no statistically significant difference between the removals at the various time periods ($F<F_{critical}$, $p>0.05$), with the exception of geosmin removals, in particular for sand 6.5 and filter coal 1.2-1.4 at 10 months, where the removals appeared to decrease. This suggests that there was no improvement in the removals of these compounds during the trial. Similarly an analysis of the data showed there was no significant decrease in the removal by the GAC over the trial period. As the percent removals for the two inlet concentrations (low and high) for each of the spiking time periods were within experimental error all data for the study were combined (Figure 16 and Figure 18).

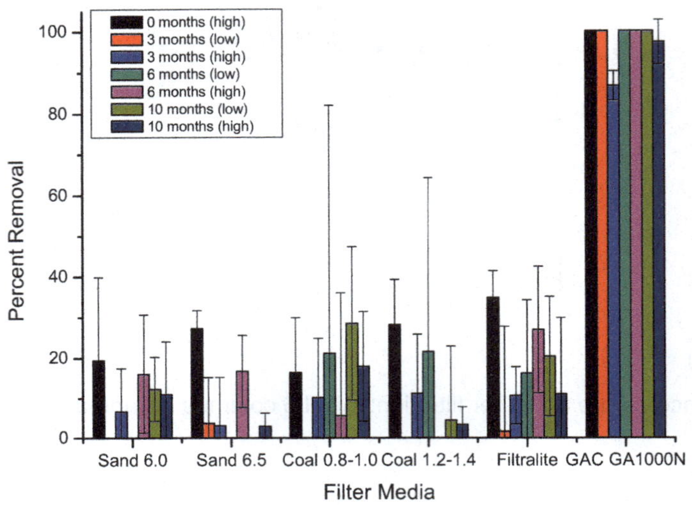

Figure 15: Percent removal of MIB by six filter media at 0, 3, 6 and 10 months after commissioning; spiking was undertaken at two concentrations (high and low, see Table 3 for details). Zero or negative values not shown.

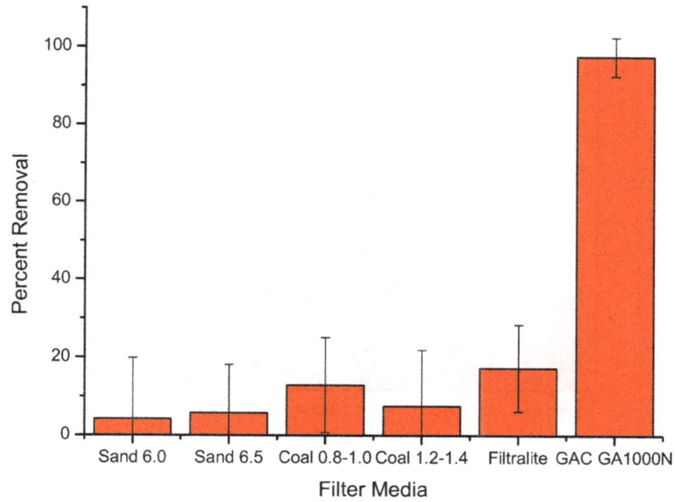

Figure 16: Average percent removal of MIB over 10 month period.

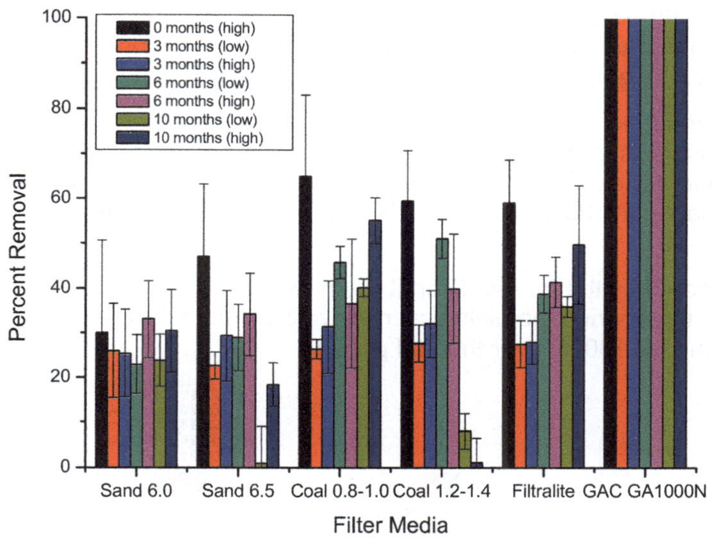

Figure 17: Percent removal of geosmin by six filter media at 0, 3, 6 and 10 months after commissioning. Zero or negative values not shown.

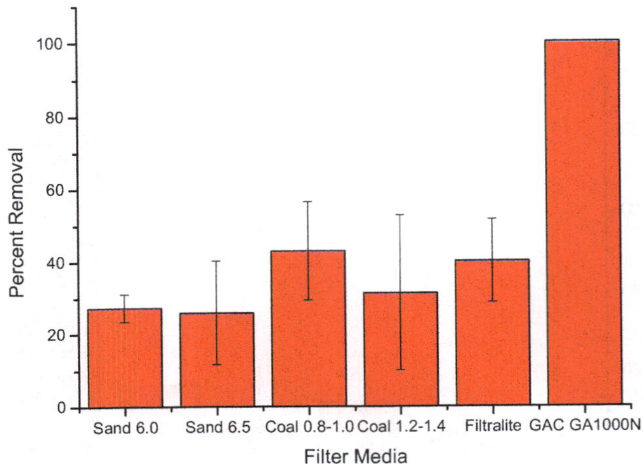

Figure 18: Average percent removal of geosmin over 10 month period.

GAC was very effective for the removal of MIB and geosmin over the trial period with an average of 98 and 100% removal respectively. This result supports previous work by Ho (2004) who investigated the removal of MIB and geosmin by GAC at the laboratory and pilot scale. Such high removals had been shown previously to be attributable to physical adsorption only (Ho and Newcombe 2010); however as the removals were high throughout the test period it was not possible to differentiate between physical adsorption and biological degradation in this study.

The lower removals and higher variability observed by the other filters are attributed to physical losses through volatilisation and adsorption to the apparatus. Previous research has consistently shown that system losses are higher for geosmin than for MIB, and are typically in the range of 30-40% and 10-20%, respectively (Elhadi et al. 2004; Ho et al. 2007a). These results suggest that no biological removal had occurred.

The results were replicated by the four additional columns (Figure 19-Figure 22); the smaller sand particle did not appear to enhance the biological removals of the metabolites, and the F300 GAC performed as well as the GA1000N over the trial period.

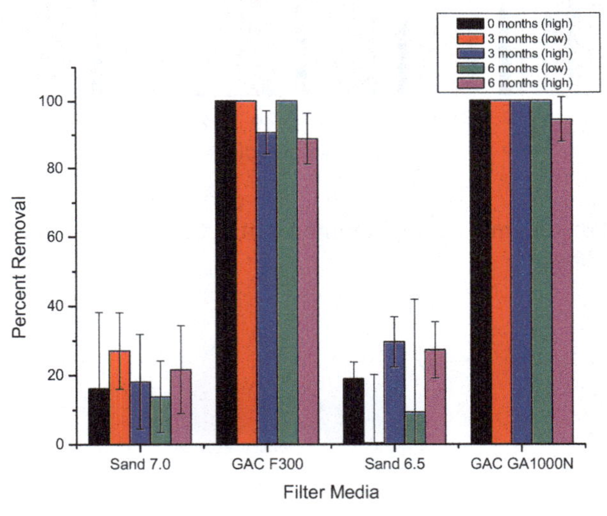

Figure 19: Percent removal of MIB by four filter media at 0, 3 and 6 months after commissioning.

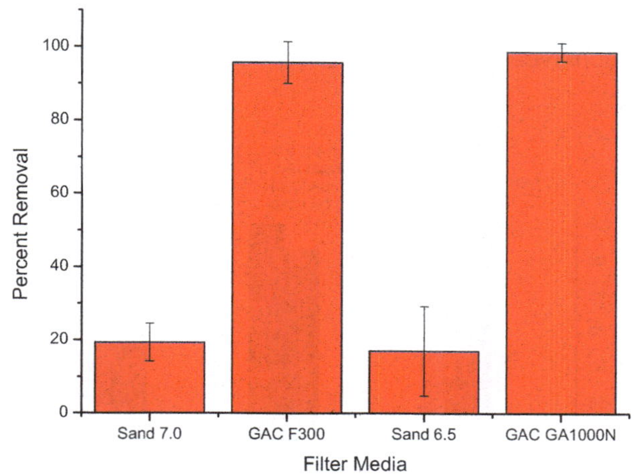

Figure 20: Average percent removal of MIB over 6 month period.

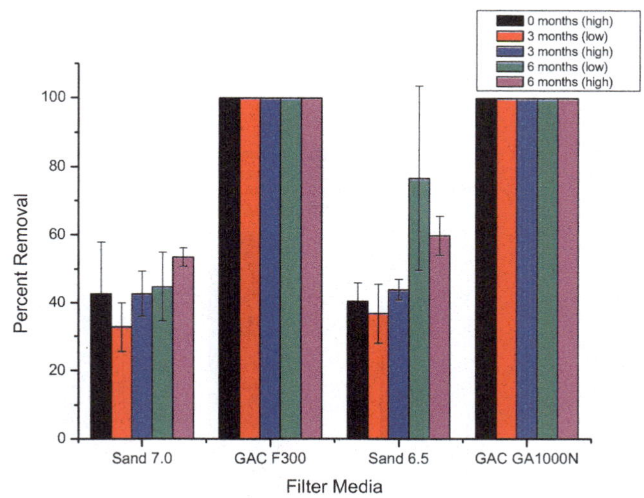

Figure 21: Percent removal of geosmin by four filter media at 0, 3 and 6 months after commissioning.

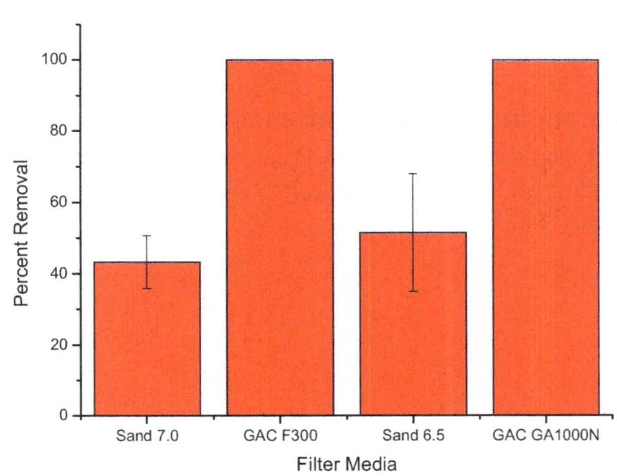

Figure 22: Average percent removal of geosmin over 6 month period.

The results for the duplicate columns over six months are shown for MIB and geosmin in Figure 23 and Figure 24 respectively. The results demonstrate good reproducibility of the column data over the trial; although there is some evidence that the removal of geosmin on filter sand 6.5 2 may have been increasing and could have achieved biological filtration in less than 10 months. Unfortunately this test was terminated at 6 months.

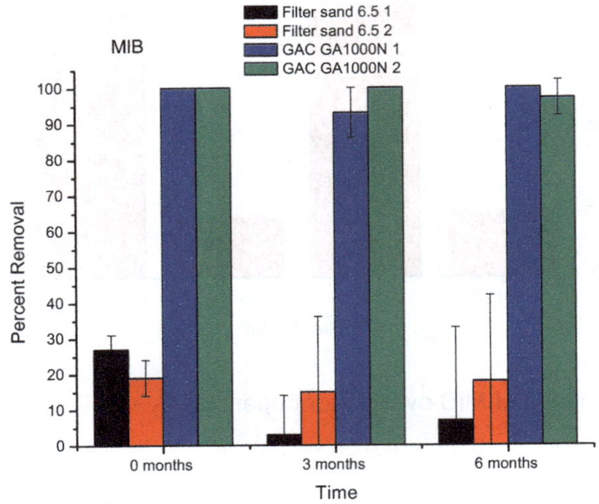

Figure 23: Duplicate column results, MIB.

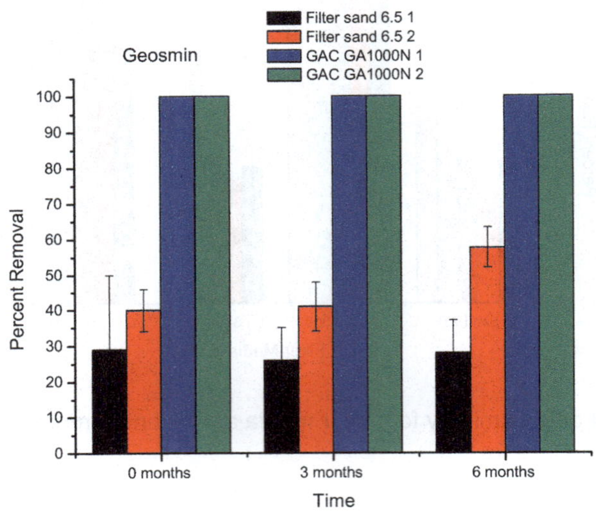

Figure 24: Duplicate column results, geosmin.

3.1.1.2.2 Cylindrospermopsin, saxitoxin and microcystin

Similar to MIB and geosmin, the removals of cylindrospermopsin, saxitoxin and microcystin were low through the non-porous media, and significant through GAC. The datasets for the spiking periods for the initial six columns are shown in Figure 25.

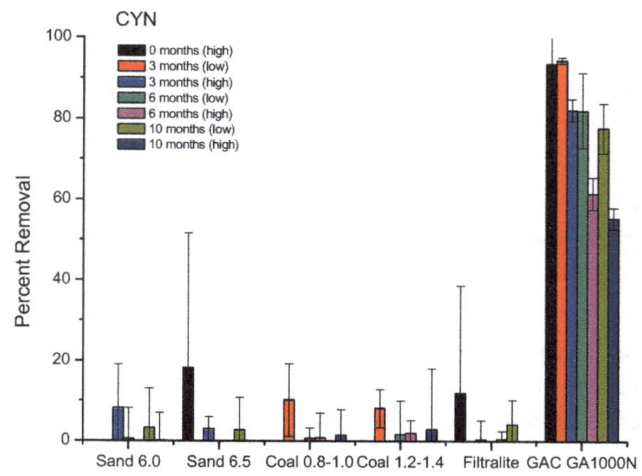

(a)

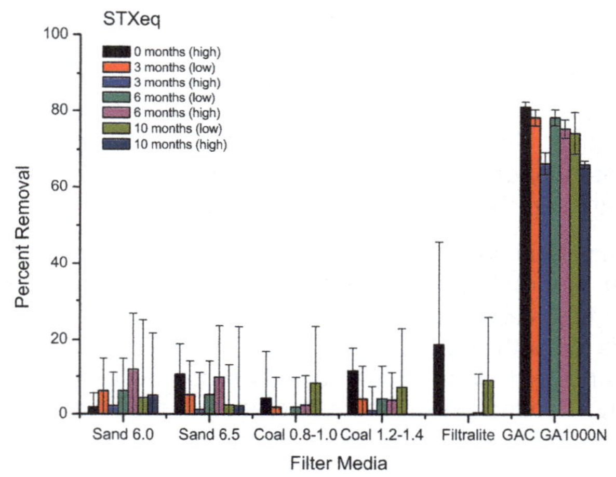

(b)

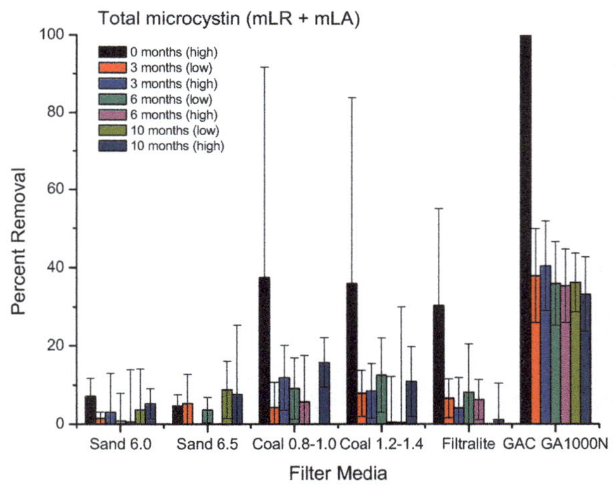

(c)

Figure 25: Percent removal by six filter media at 0, 3 6 and 10 months after commissioning for CYN (a), STXeq (b) and microcystin (total) (c). Zero or negative values not shown.

Although there was some variability in the removal through the non-porous media, the removals were very low or zero. As all positive values had standard deviation errors that included zero, apparent removal was attributed to physical adsorption onto apparatus and media surfaces and/or biofilm. The results for the GAC suggest a decreasing trend in the removal of total microcystin and cylindrospermopsin with time. Similar removal trends were shown by the additional four columns (Figure 26).

The results for the GAC over the two trials are compared in Figure 27. As previous work has shown that the adsorption of mLA is lower than mLR (Newcombe et al. 2003; Wang et al. 2007) these two toxins were separated for these graphs (c) and (d). Where columns are not visible samples were not taken.

At commissioning the removal of cylindrospermopsin was close to 100%, in agreement with previous data (Ho et al. 2011). Although the removal at 10 months was still high for GA1000N (average 67%), there appears to be a decrease with time. For this metabolite GA1000N is slightly more effective than F300 at 3 and 6 months.

The removal of STXeq remains relatively stable over the trials, within experimental error, at approximately 76%. Previous studies have shown that the removal of some saxitoxin variants decreases significantly with time, due to the adsorption of natural organic carbon molecules reducing the adsorption sites available for the metabolites (Newcombe 2002; Newcombe and Nicholson 2004). In those studies individual variant concentrations were measured and the overall toxicity was calculated as described by Newcombe (2002). In the present study the analytical technique (ELISA) gives only an estimate of the overall toxicity (as STXeq) and the result is dependent on the relative concentrations of the variants. The relatively constant removal shown in Figure 27 (b) is probably a reflection of the lower sensitivity of the analysis to the less toxic variants which also show the greatest decrease in adsorption with time (Newcombe and Nicholson 2002).

Comparison of microcystin variants showed that F300 was more effective for the removal of these toxins than GA1000N, in contrast to the adsorption of cylindrospermopsin. Figure 27 (c) and (d), shows that mLA is less effectively removed by GAC than mLR after commissioning and there is a decrease in the removals of both toxins within the first 3 months which is in agreement with data reported by Newcombe and Nicholson (2004) and Newcombe (2002) for several pilot trials lasting over 12 months. The authors found a significant decrease in the removal of mLA and mLR within one to three months of commissioning of the filters. In some studies, such a decrease in adsorption with time has been followed by an increase in removal due to the onset of biological degradation (Newcombe et al. 2003; Wang et al. 2009). Biodegradation did not commence during this 10 month trial as is evidenced by a sharp increase in the removal (Newcombe et al. 2003; Wang et al. 2007)

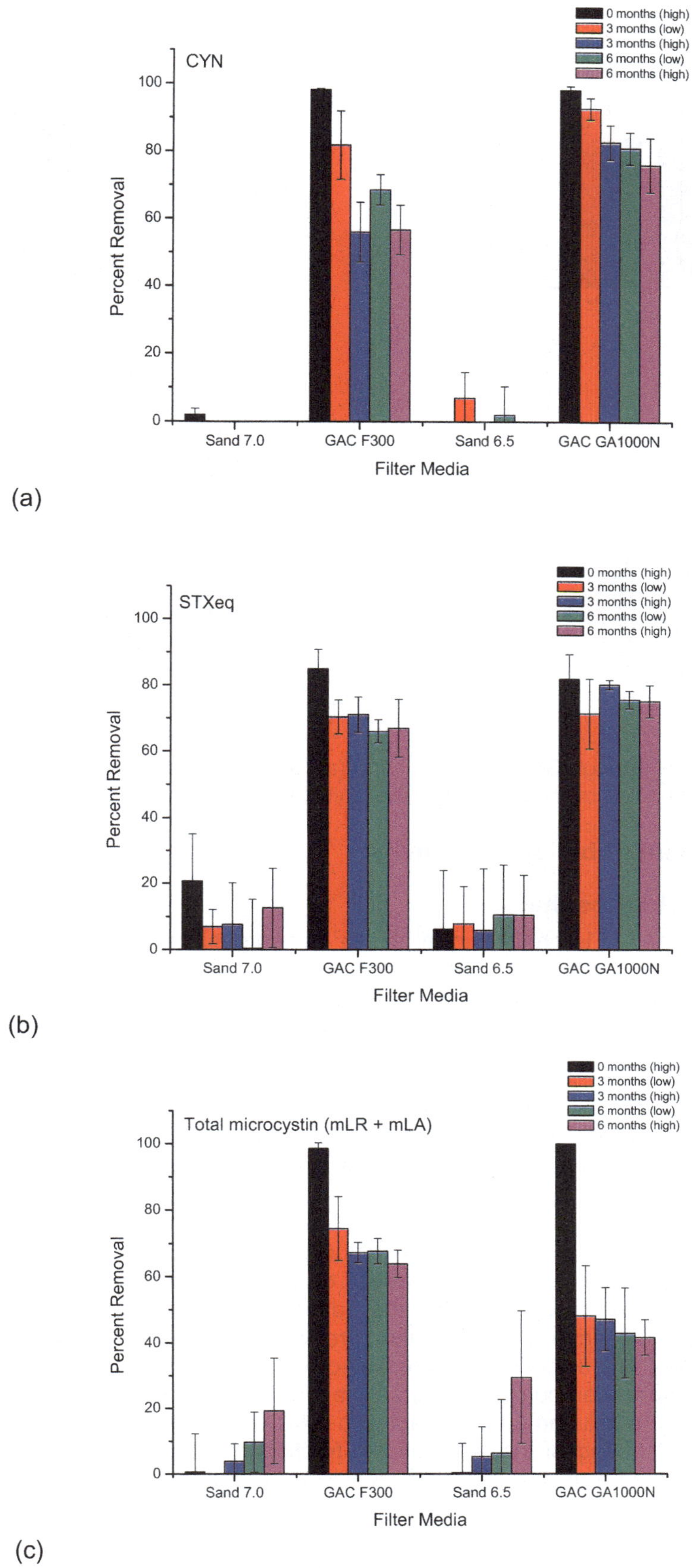

Figure 26: Percent removal by additional four filter media at 0, 3 and 6 months after commissioning for CYN (a), STXeq (b) and microcystin (total) (c). Zero or negative values not shown.

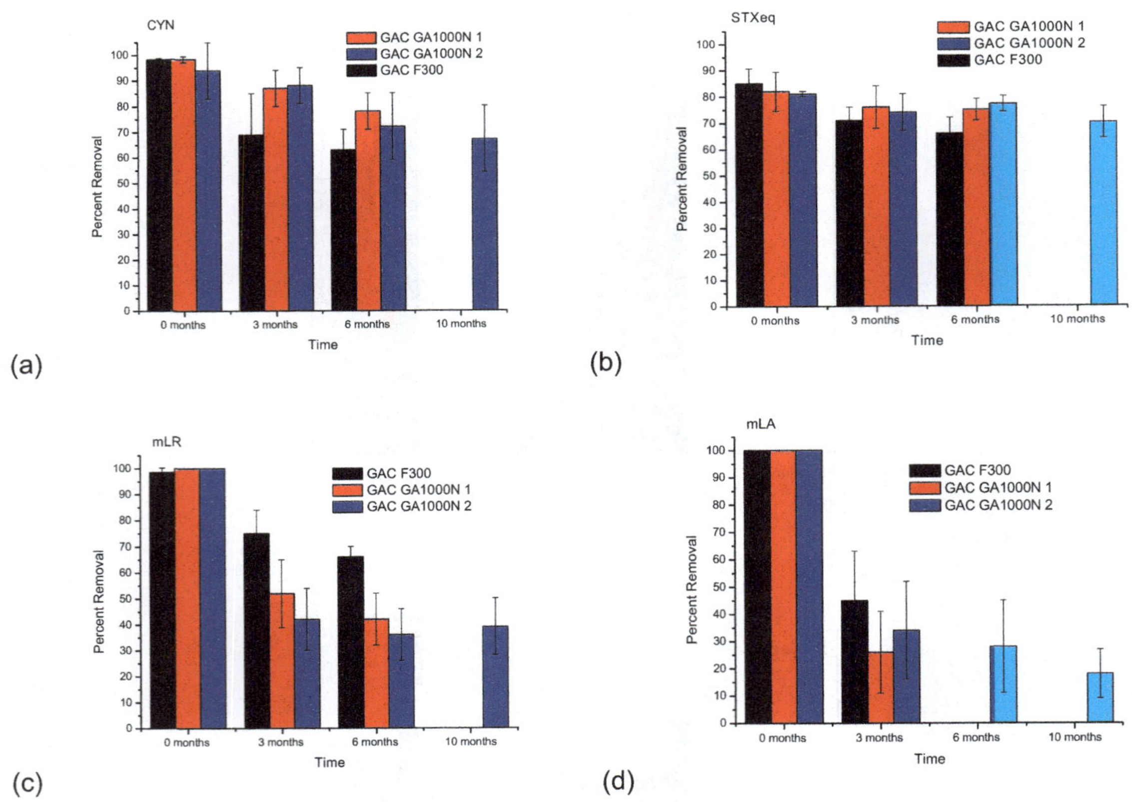

Figure 27: Percent removal through GAC media for CYN (a), STXeq (b), mLR (c) and mLA (d), mLA at time 6 months percent removal not shown due to below the limit of reporting.

3.1.1.3 Measurement of biological activity

3.1.1.3.1 Adenosine triphosphate

Adenosine triphosphate (ATP) is present in all active microorganisms therefore ATP concentrations can be used as a measure of biological activity and an indicator for the presence of a biofilm on filter media. Samples of the filter media were taken prior to the spiking trials and directly after collecting the final effluent samples from the spiking trial to look at the impact of the spiked settled water on the biological activity of the filter media. From Figure 28 it can be seen that there was measurable ATP on the sand, filter coal and Filtralite media, indicating the presence of a biofilm. However, the results showed no clear trend. After the spiking trials there appears to be higher ATP present and hence biological activity for the majority of the filter media. In contrast to the other media the GAC appeared to have negligible ATP activity before and after the spiking trials (as shown in shaded region of Figure 28).

This result was unexpected based on previous results (Velten et al. 2007) and GAC surface is known to promote significant biofilm growth within 4-6 weeks of filter commissioning (Newcombe 2002). The low value of ATP on the GAC suggested that the adsorbent interfered in some way with the ATP analysis. To determine the extent of this effect ultrapure water was spiked with a known concentration of ATP. Three samples of this solution were contacted with GAC, and the ATP concentrations were measured. The results were then compared with a standard curve prepared in the absence of GAC. The concentration of ATP in solution after contact with GAC was 0.1% of the expected concentration. The results from this study suggest that a significant fraction of the ATP released during lysis of the biofilm on the GAC sampled from the column was adsorbed and thus was not measured using the normal ATP analytical procedure. Most literature suggests that this ATP method is suitable for GAC (Velten et al. 2007). The limited literature available on the effect described above suggested that an alternative method, of vortexing the filter medium to remove the biofilm prior to ATP analysis, required optimisation with the particular filter medium and biofilm (Magic-Knezev and Kooij 2004). Due to the time constraints an alternative method (scanning electron microscopy) was used to qualitatively determine the presence of a biofilm on the various filtration media.

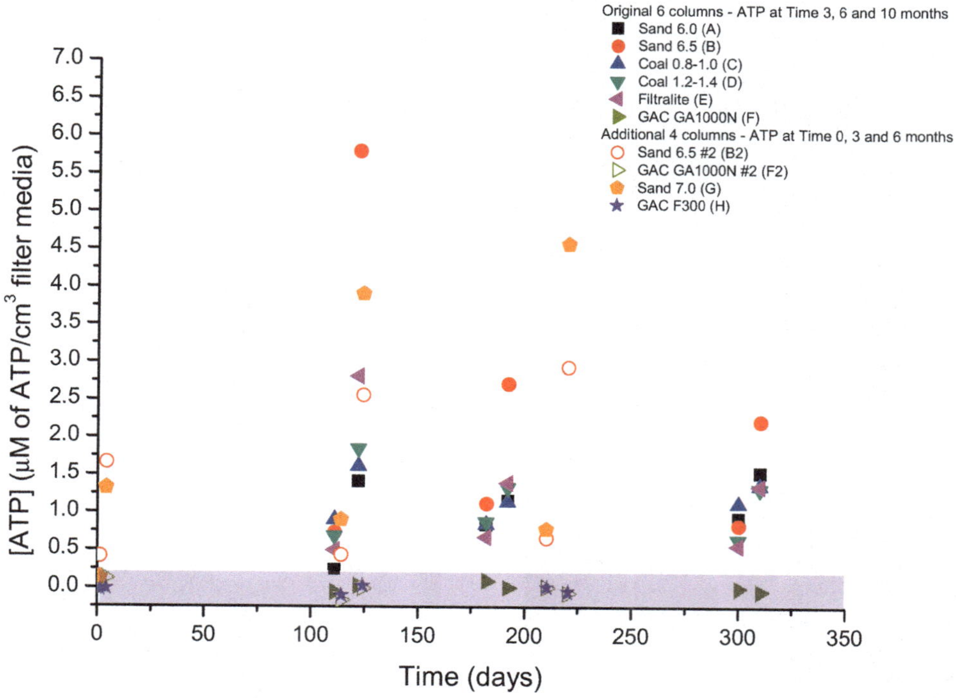

Figure 28: ATP concentrations during the laboratory filter trials. Shaded area highlights GAC ATP activity.

3.1.1.3.2 Scanning electron microscopy

Scanning electron microscopy (SEM) is a qualitative method that can be used to investigate the existence and composition of a biofilm on filter media. Although SEM does not allow the identification of individual organisms it can be used to show the abundance and diversity of organisms within the biofilm.

Figure 29 to Figure 34 show scanning electron micrographs of the initial six filtration media (a) as new, prior to the column trials and (b) after 6 months in contact with Happy Valley settled water. For all media it is clear that a significant biomass has accumulated during the six month trial. The biofilms are composed of a variety of organisms, including bacteria, protozoa, fungi and diatoms and it is held together by extracellular polymeric material. It is not possible to identify individual microorganisms, and no differences can be established between the different media. However, it is clear that an abundance and diversity of microorganisms exist within the thriving biofilms.

The electron micrographs also show the differences between the surface topography of the new surface media, which might be expected to influence the formation and adhesion of a biofilm. Rough surfaces, with sheltered areas are generally considered to be more favourable for biofilm formation as it renders the biomass some protection, and thus less susceptible to disruption due to backwashing or extreme changes of flow. From the SEM analysis it appeared that all media consisted of rough and smooth surfaces; however the Filtralite and GAC displayed more extensive areas of extreme roughness than other media.

(a)

(b)

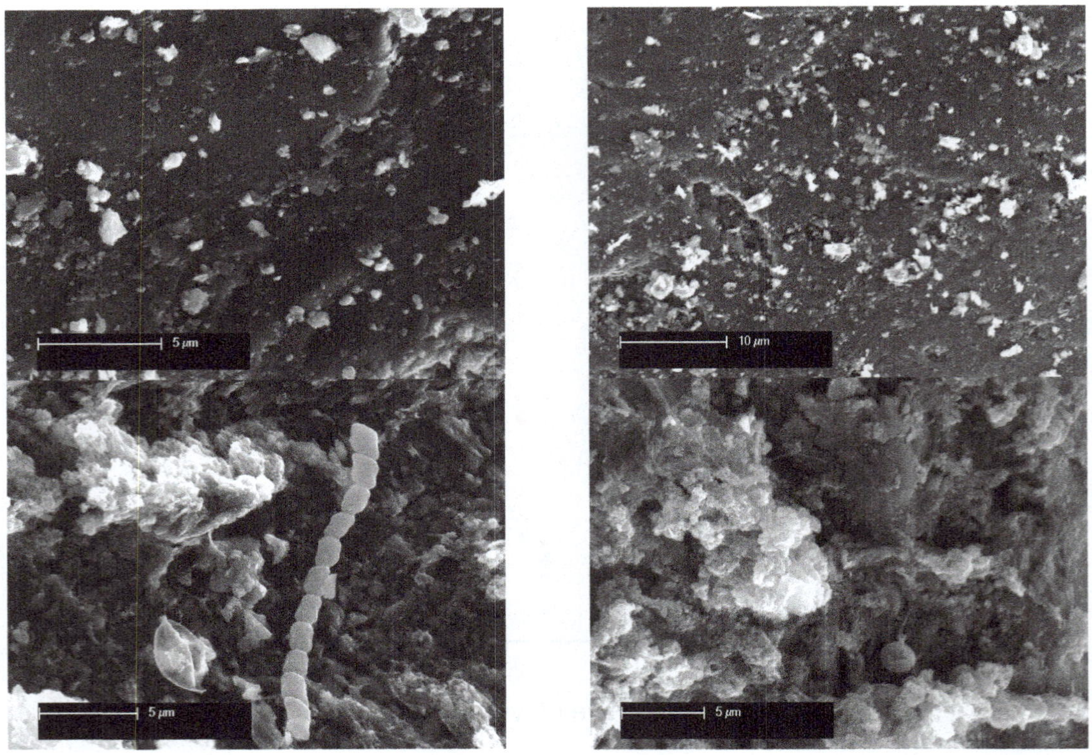

Figure 29: Sand 1.5-1.7 mm diameter, new (a) and 6 months after commissioning (b).

(a)

(b)

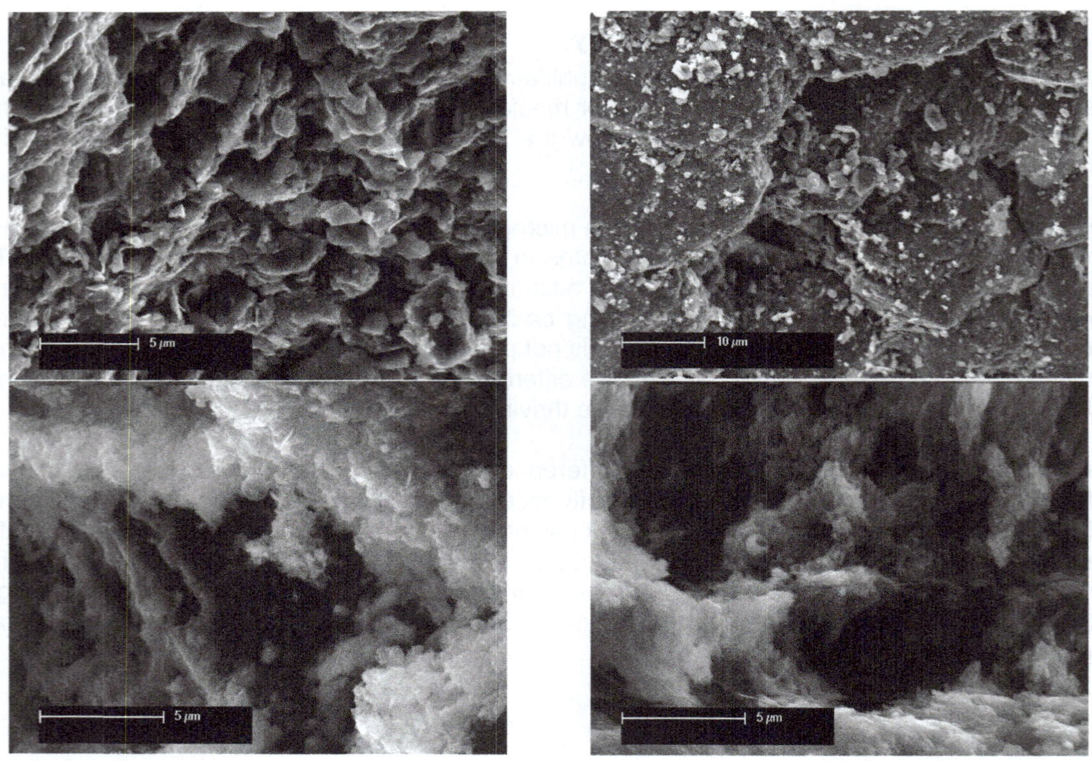

Figure 30: Sand 0.85-0.95 mm diameter, new (a) and 6 months after commissioning (b).

(a)

(b)

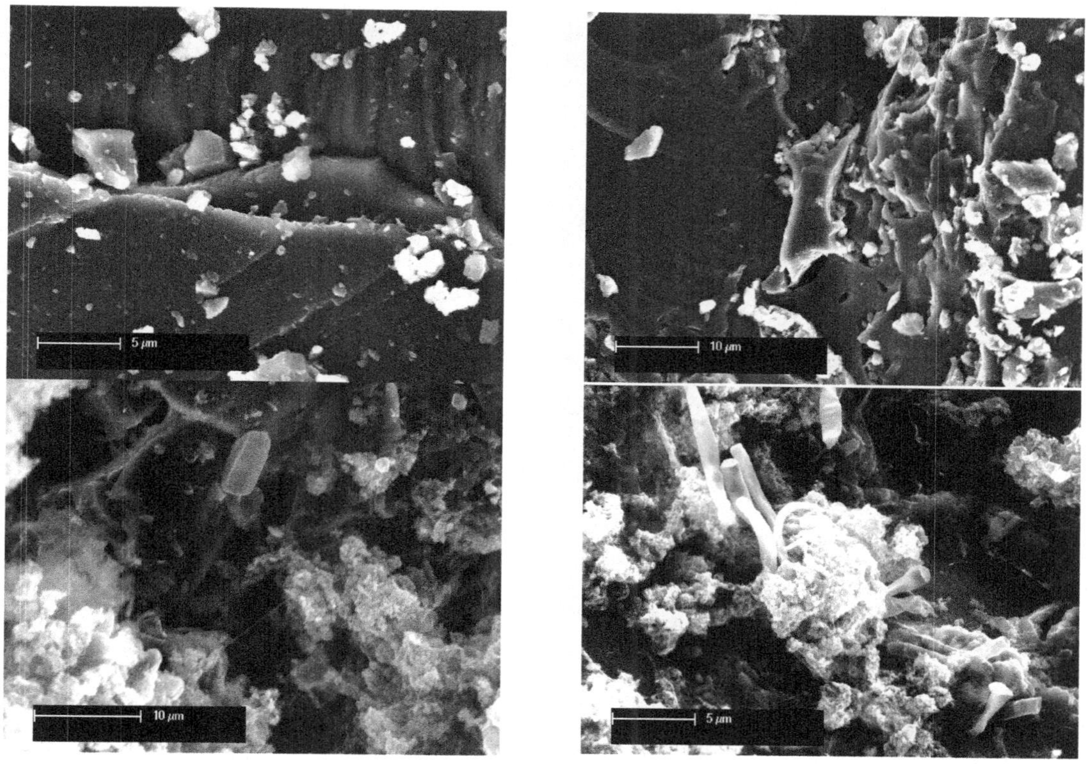

Figure 31: Filter coal 0.8-1.0 mm diameter, new (a) and 6 months after commissioning (b).

(a)

(b)

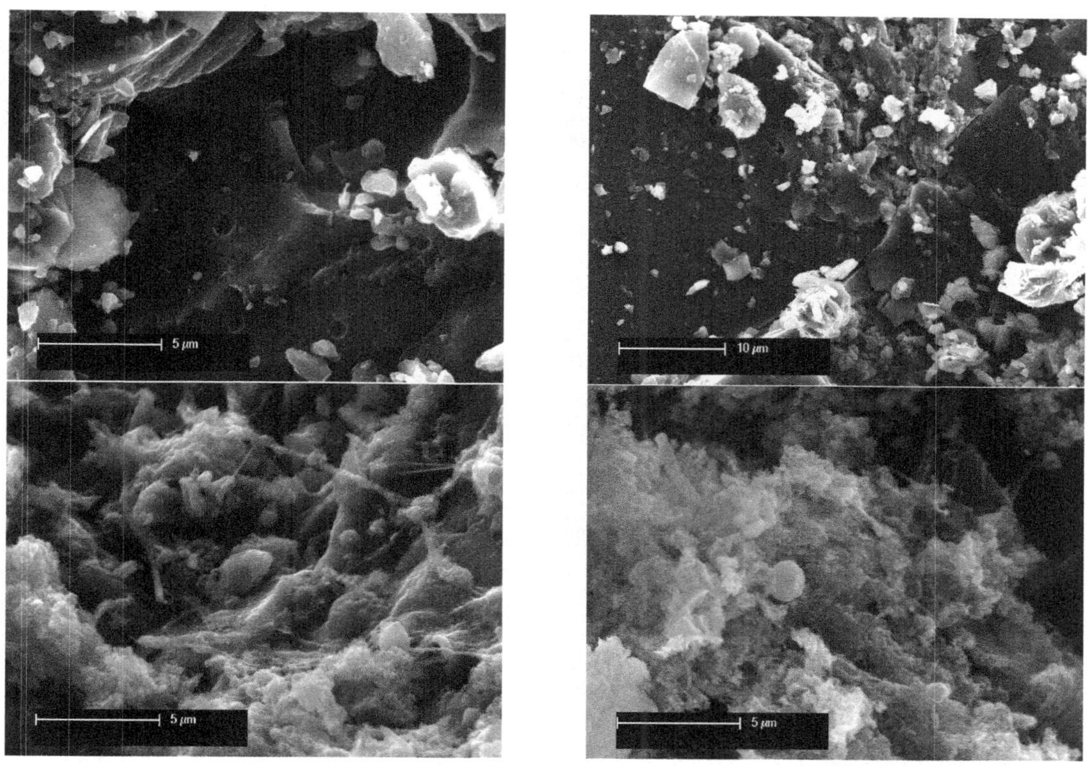

Figure 32: Filter coal 1.2-1.4 mm diameter, new (a) and 6 months after commissioning (b).

(a)

(b)

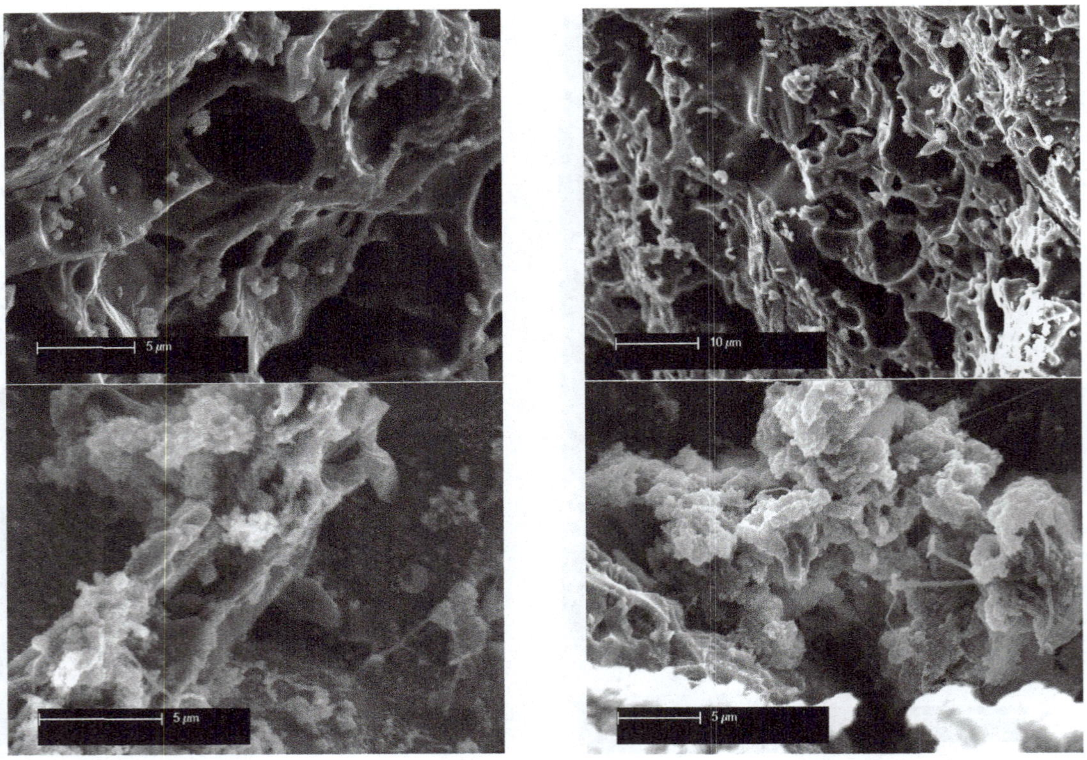

Figure 33: Filtralite (expanded clay) particle size 0.8-1.6 mm diameter, new (a) and 6 months after commissioning (b).

(a)

(b)

Figure 34: GAC (GA1000N) particle size 0.5-0.7 mm diameter, new (a) and 6 months after commissioning (b).

3.1.1.4 Summary and Conclusions

During the trial period all filtration media developed significant biomass. However, there was no evidence of the onset of biological degradation of the cyanobacterial metabolites for any of the filter media. The results suggest that none of the filter media could be expected to significantly reduce the time taken for the onset of degradation compared with the preliminary work shown in Figure 3. The non and slightly porous media (sand, filter coal and Filtralite) displayed removals of the metabolites consistent with physical removals, e.g. volatilisation and adsorption onto the apparatus. As the removals of MIB and geosmin observed on the GAC filters at the conclusion of this study were still >98% it is not possible to discriminate between physical and biological processes and these results may have been influenced by both. However, the removal of the toxins displayed a trend of reduction of removals with time, consistent with a reduction of the available adsorption sites and therefore physical removal only.

Based on this data, prior to the onset of biological degradation, GAC is the most effective filtration medium for the removal of UV absorbing compounds, DOC, taste and odour compounds MIB and geosmin and the cyanotoxins, microcystins, saxitoxins and cylindrospermopsin. This is not unexpected as GAC has a higher surface area for adsorption and can therefore remove compounds via two mechanisms, adsorption and biological degradation. Such capacity for physical adsorption during the initial phase of filter life ensures that sufficient removal of cyanobacterial metabolites is achieved while biofiltration processes are establishing on the medium surface.

In summary, the results of this study indicate that 10 months after commissioning of the filters with virgin media there was no evidence of biological removal of any of the cyanobacterial metabolites, although all media developed a biomass. While different media with various particle size and surface properties will have an effect on biofilm formation, this data suggests that organisms capable of degrading the various cyanobacterial metabolites have not yet established themselves within the biofilm, and under these conditions the smaller particle size, rougher surface topography and different surface chemistry did not significantly enhance the development of a *degrading* biomass.

3.1.2 Effect of temperature on biodegradation of cylindrospermopsin and saxitoxins

Figure 35 and Figure 36 show the results of batch biological degradation experiments for STXeq, in Happy Valley and Swan Reach settled water and CYN, in Myponga floated and Swan Reach settled water, at four temperatures respectively.

For STXeq, there is a decrease in concentration over 190 days at 20°C and 30°C (approximately 50%) compared with 5°C and 15°C in both the sterile and non-sterile reactors. There also appears to be an increase in the overall toxicity of the water for several samples that may indicate that some of the variants have undergone transformation to more toxic variants. Similar results were seen for both waters, Myponga and Swan Reach (Figure 35). Although these results suggest complex effects on the toxicity of the water measured by ELISA, it appears these may be a function of temperature rather than biological activity as sterile and non-sterile samples displayed similar behaviour.

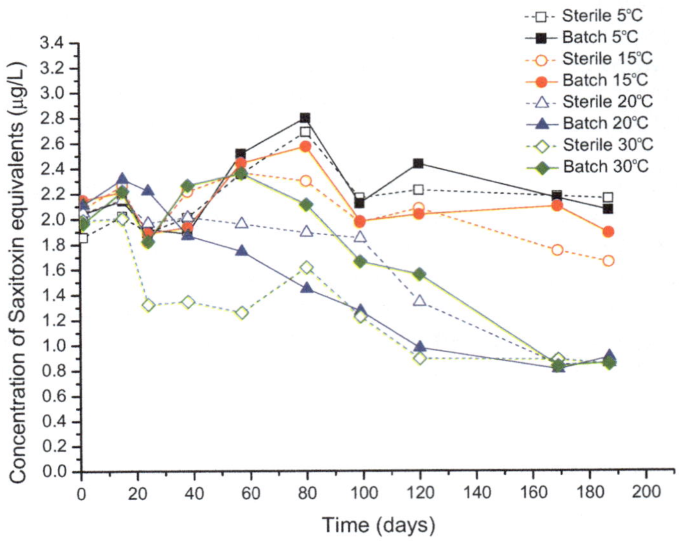

(a)

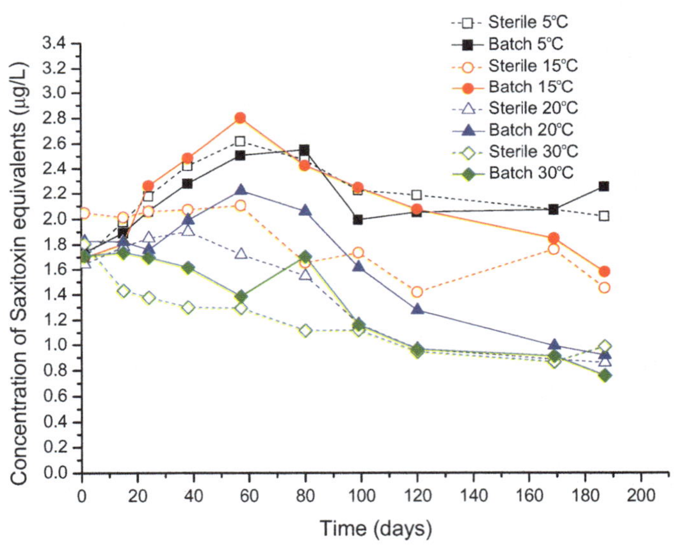

(b)

Figure 35: Concentration of STXeq over time in batch experiments, Happy Valley (a) and Swan Reach (b) settled water.

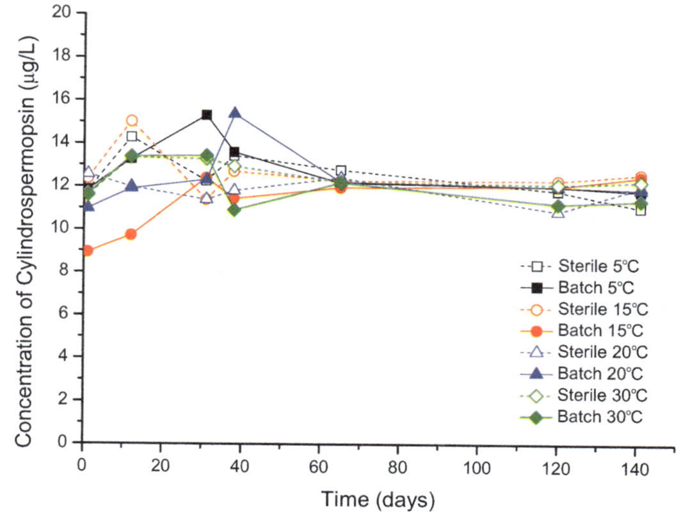

(a)

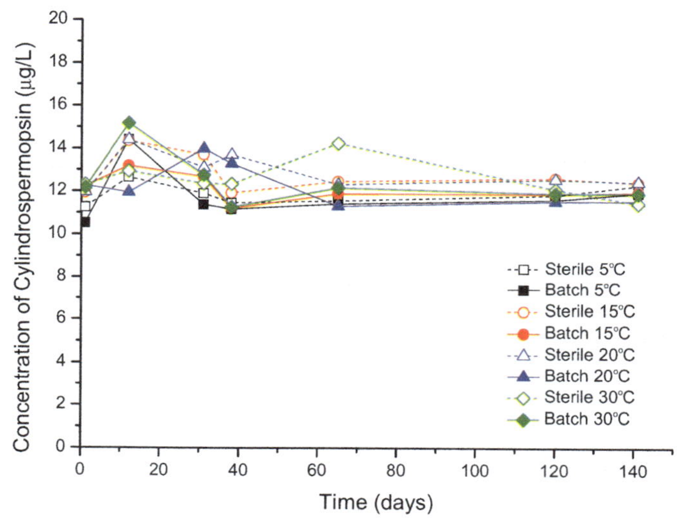

(b)

Figure 36: Concentration of cylindrospermopsin over time in batch experiments, Myponga floated (a), Swan Reach settled (b) water.

After over 140 days, the batch experiments for CYN in Myponga and Swan Reach waters displayed similar results, as shown in Figure 36, with no degradation observed in either water. The small variance within results in comparison to the original concentration spiked, 10 µg/L, is within experimental error.

It was thought that the Swan Reach water may display some degradation over this time period as in more recent years the Murray River in South Australia has experienced several blooms of *Cylindrospermopsis raciborskii*. Although toxin had not been detected in the Murray River samples, it was considered possible that toxins had been produced at some stage.

Additional batch experiments were undertaken at 15 and 20°C comparing Swan Reach raw (Murray River) water and settled water sampled at the treatment plant at the same time. These results are shown in Figure 37. No degradation was observed in the settled water at either temperature which is in agreement with the previous batch experiments (Figure 36). Similar results were observed for the raw water at 15°C; however by increasing the temperature to 20°C it was observed that CYN degraded within 18 days to below the limit of detection.

The results of these batch experiments suggests that bacteria capable of degrading CYN are present in Murray River raw water at Swan Reach but are removed by coagulation processes at the water treatment plant and the bacteria degrade CYN more effectively at 20°C than at 15°C.

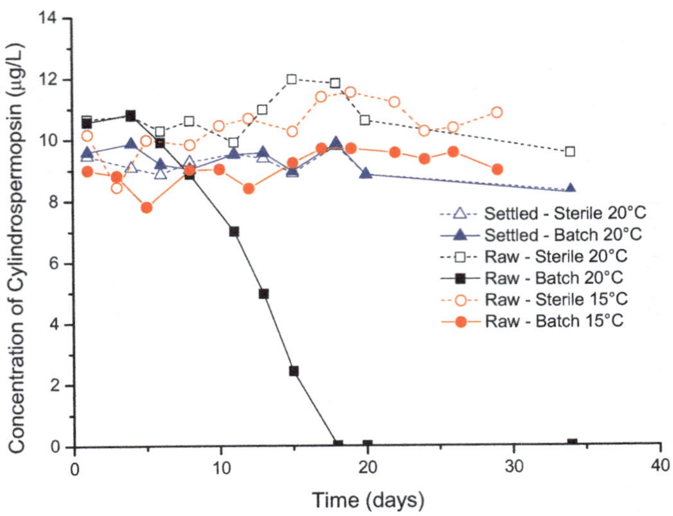

Figure 37: Results of cylindrospermopsin batch experiments, Swan Reach.

3.1.3 Effect of hydraulic loading on biological degradation of MIB and geosmin

The effect of increased flow on removal of MIB and geosmin through sand filters with an existing biofilm was determined using a pilot plant commissioned at Morgan WTP. Four EBCTs were studied, 5, 8, 10 and 15 minutes, each over a 1 week period. MIB and geosmin were spiked in at a target concentration of 100 ng/L, actual MIB concentrations were 81 ng/L ± 10 ng/L and geosmin 131 ± 24 ng/L.

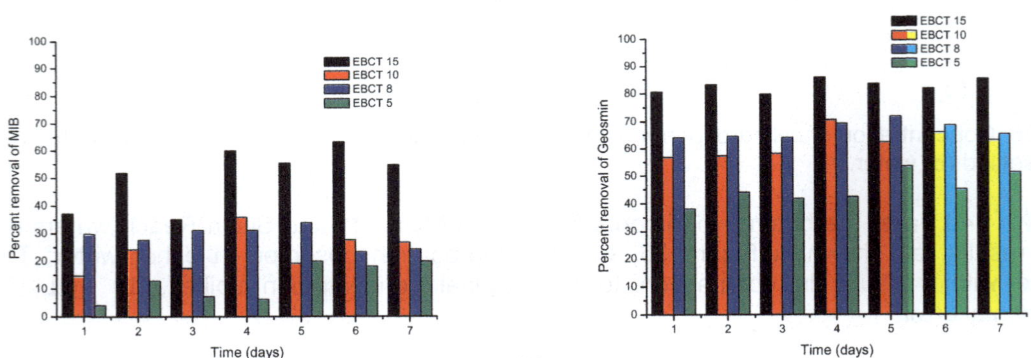

(a) (b)

Figure 38: Comparison of removals for 4 EBCTs, MIB (a) and geosmin (b).

Figure 38 shows the percent removal of MIB (a) and geosmin (b) at each EBCT over the week-long spiking period. At the highest EBCT (15 minutes) the filters display the highest removals (average 55% and 80% for MIB and geosmin respectively). While both values are higher than generally found for physical removals, they were not as high as those found in previous laboratory testing of Morgan sand at the same EBCT where 100% removal for both T&O was observed (McDowall 2008). The full-scale Morgan filters have been recognised as efficient biological filters for the removal of geosmin over many years (Figure 4). The removals at the equivalent EBCT (<5 min) in this study were more consistent with physical processes only.

Since 2011 MIB has intermittently also been an issue for Morgan WTP, with concentrations of up to 64 ng/L entering the plant. Removals at the full scale in 2011 and 2012 were also lower than predicted from the previous laboratory-scale columns, and more consistent with these pilot studies (conducted in 2011/2012).

From these results it appears that biological degradation may be responsible for the observed MIB and geosmin removals at the EBCT of 15 minutes, and for geosmin at the 10 and 8 minute EBCTs. Higher removals of T&Os were observed from higher EBCTs, suggesting hydraulic loading can have an effect on the removal rate of these metabolites on both physical and biological processes. The differences observed between the full-scale removal of geosmin and at the pilot scale may be a result of the media. In the pilot columns only sand was used whereas the Morgan filters are sand/anthracite. It is possible that higher biological removals are experienced within the biofilm on the anthracite and this effect would not have been observed in this study.

3.2 Biological Filtration Potential Test

The results for cyanobacterial metabolite removal for two of the WTPs tested (SA Water and Hunter Water) are shown in the following sections; water and filter media from the other three plants showed no differences between sterile and non-sterile samples for the degradation of metabolites and are not shown. Results for those tests are given in Appendix II.

3.2.1 SA Water, Morgan Water Treatment Plant

3.2.1.1 Laboratory-scale filtration

The results of column testing to identify biological removal of MIB, geosmin, STXeq and mLR during filtration through sterile and non-sterile sand from the Morgan Water Treatment Plant (WTP) are shown in Figure 39. The influent water was sampled from the settled water ducts at the plant.

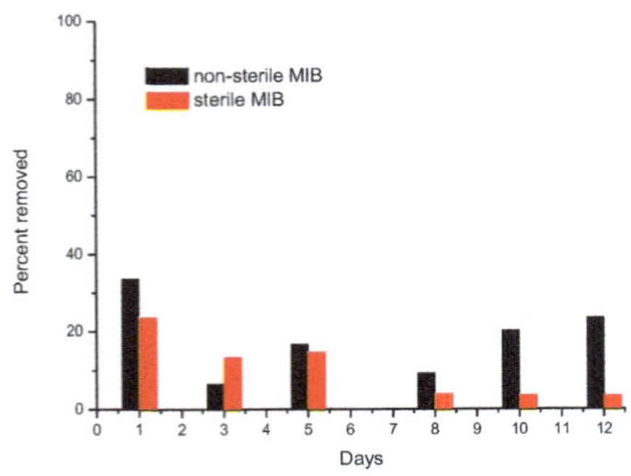

(a)

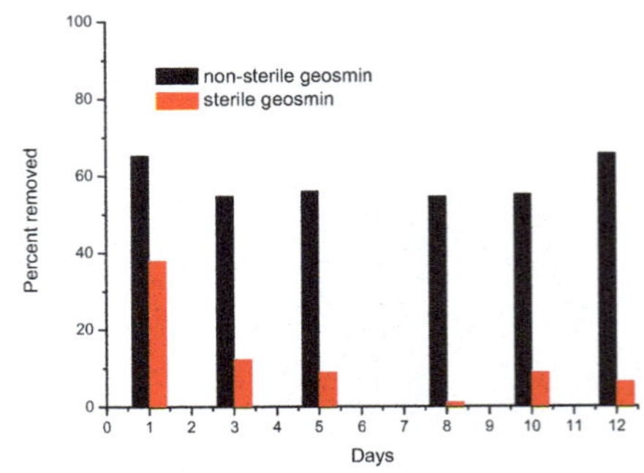

(b)

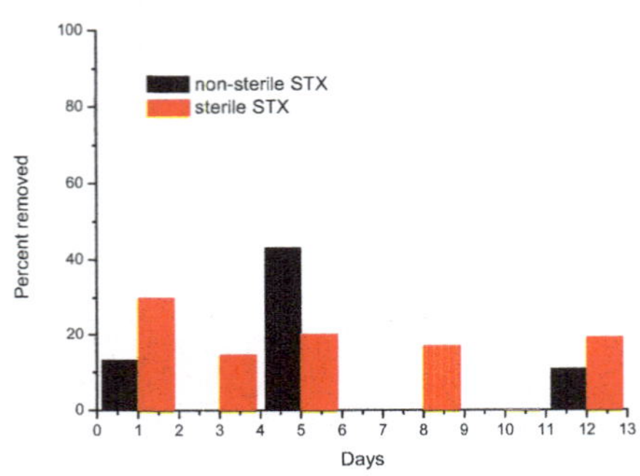

(c)

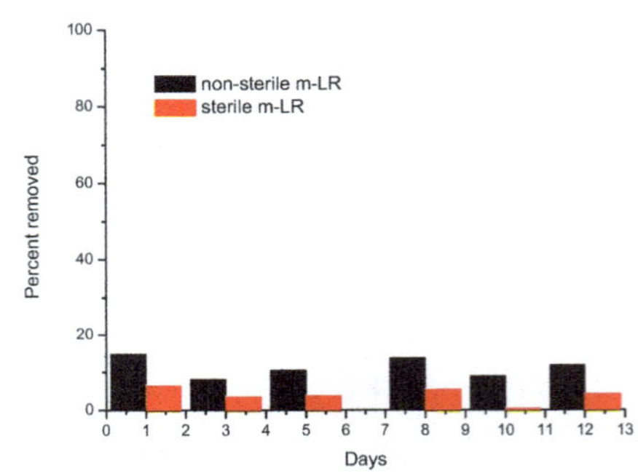

(d)

Figure 39: Column experiments, percent removals over 12 days: MIB, initial concentration 48 ng/L (a); geosmin, initial concentration 34 ng/L (b); saxitoxin equivalents, initial concentration 2.6 µg/L (c); microcystin LR, initial concentration 11 µg/L (d). Zero and negative results not shown.

Figure 39 (a) suggests no biological removal of MIB as there is no significant difference between the sterile and non-sterile columns and the percent removals are more consistent with physical processes. In contrast, the large difference between the sterile and non-sterile sand for geosmin indicates biological removal and is consistent with the results of the pilot trial. However, the results are significantly different from previous laboratory studies which showed biological removal of both compounds to below detection under the same conditions. These results suggest that the biofilm present in the Morgan filters has been impacted in some way to reduce the MIB removal potential.

The results for STXeq and mLR suggest no biological removal during filtration for these metabolites.

3.2.1.2 Batch experiments, settled water

Settled water from Morgan Water Treatment Plant was spiked with MIB, geosmin, STXeq, mLR and mLA and cylindrospermopsin. Another water sample was sterilised prior to spiking. Any decrease in concentration of the non-sterile sample indicates the presence of degrading organisms in the influent water to the filters at the full scale, and the potential of these organisms to colonise subsequent filtration media.

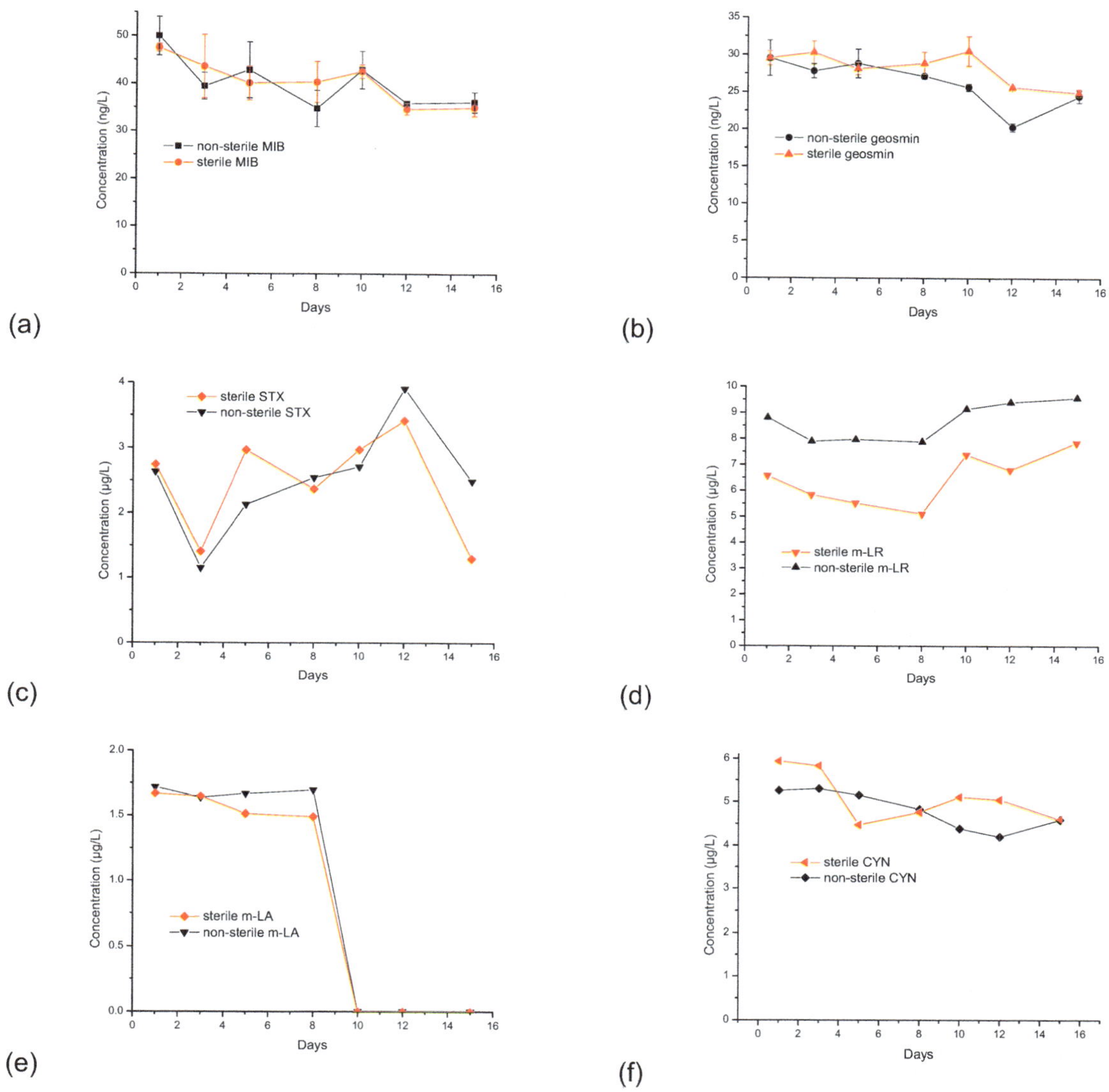

Figure 40: Batch experiments in settled water, concentration as a function of time: MIB (a); geosmin (b); STXeq (c); microcystin LR (d); microcystin LA (e); cylindrospermopsin (f).

No degradation of MIB or geosmin occurred in the batch experiments (Figure 40), indicating the absence of degrading organisms in the influent to the filters. If the present conditions no longer provide these organisms for re-seeding the filters this may result in a reduction of the efficiency of the biological filtration process. Although the concentration of mLA was low, it is clear that degradation of the toxin occurred in both the sterile and non-sterile sample, possibly due to some microbiological cross-contamination. There was no significant difference between the sterile and non-sterile reactors for the other toxins, indicating an absence of degrading bacteria.

3.2.1.3 Batch experiments, raw water

As the results for the settled water were unexpected, additional samples were taken of the raw water drawn from the Murray River, and the same batch testing was undertaken (Figure 41).

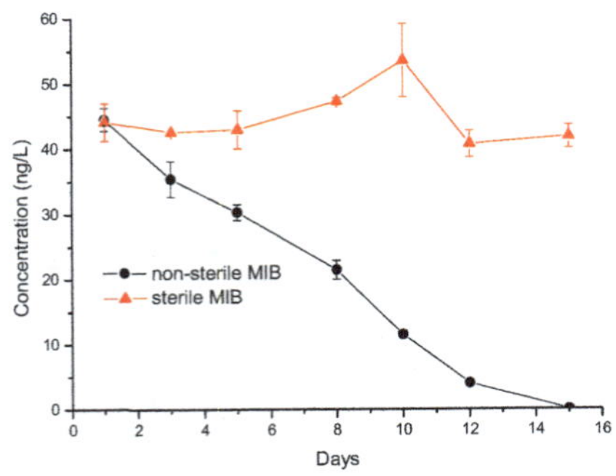

(a)

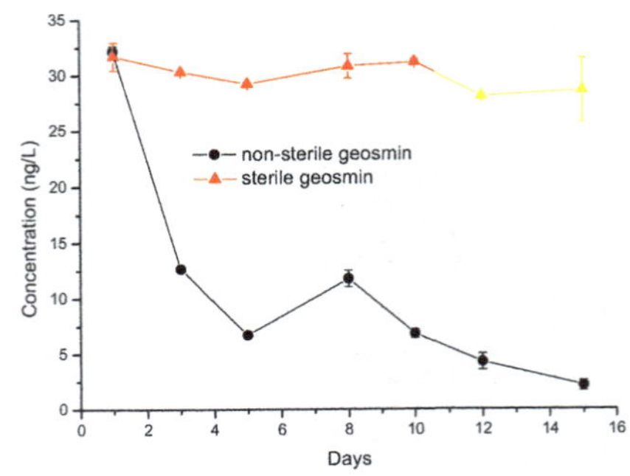

(b)

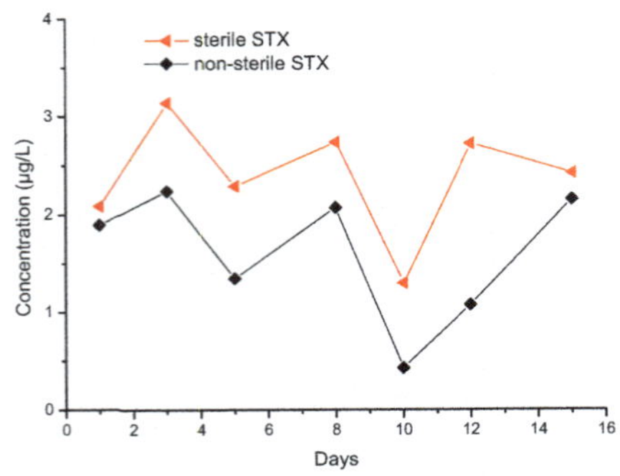

(c)

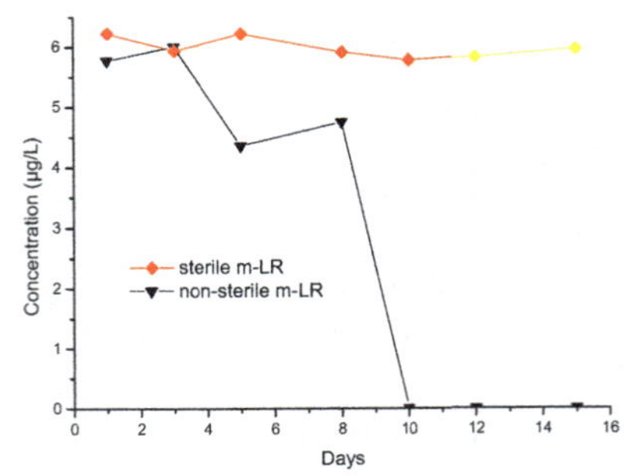

(d)

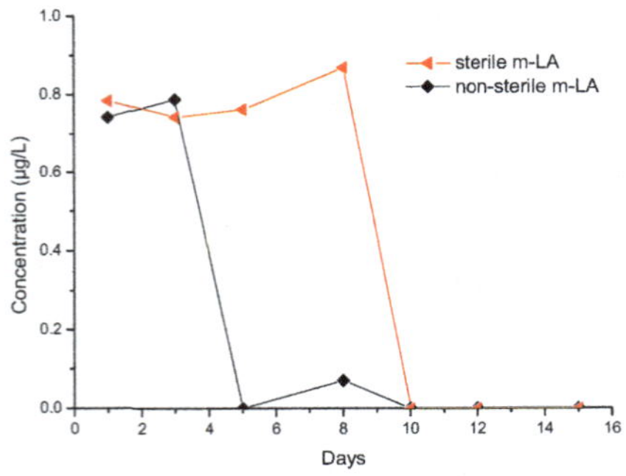

(e)

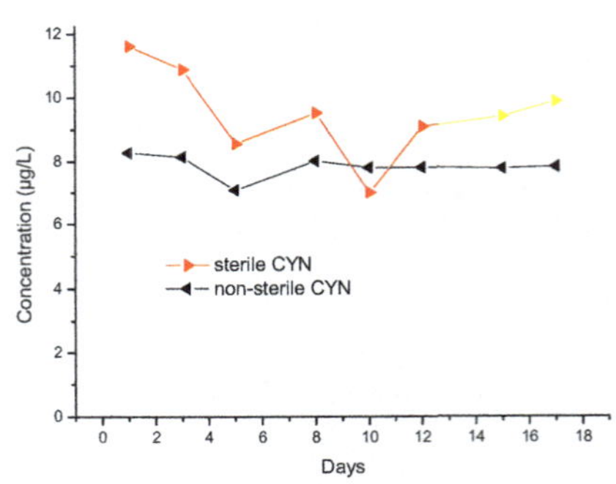

(f)

Figure 41: Batch experiments in raw Murray River water, concentration as a function of time: MIB (a); geosmin (b); STXeq (c); microcystin LR (d); microcystin LA (e); cylindrospermopsin (f).

Figure 41 shows that organisms capable of degrading MIB, geosmin, mLR and mLA were present in the inlet water to this plant. The present treatment operations were possibly removing/inactivating these bacteria, and more investigations are required to determine the cause. As with Figure 40, the results indicate the presence of mLA degrading organisms in both the sterile and non-sterile batches; it is possible that the organism is present in the microcystin spiking material, although this was not investigated further.

3.2.1.4 Summary

Historically the filters at Morgan WTP have performed efficiently for the biological removal of MIB and geosmin both at the full scale and during laboratory-scale column experiments. During this work, application of the BFP test identified some biological removal of geosmin but no significant removal of the other metabolites, in contrast to previous studies. The batch experiments confirmed the absence of MIB and geosmin degrading bacteria in the influent to the sand filters, possibly preventing the re-seeding of the degrading bacteria. Further batch experiments on the raw water confirmed the presence of bacteria capable of degrading MIB, geosmin, mLR and mLA. It is possible that present operations at the plant are preventing viable degrading bacteria reaching and re-seeding the filters. The laboratory-scale results were reflected at the full scale during a recent episode of MIB entering the Morgan WTP. Biological filtration proved ineffective for removal of MIB, resulting in the application of PAC to manage the odour episode.

3.2.2 Hunter Water, Grahamstown Water Treatment Plant

3.2.2.1 Laboratory-scale filtration

Figure 42 shows the results of the column testing to identify biological removal of MIB, geosmin, STXeq and mLR during filtration through sterile and non-sterile sand from the Grahamstown WTP, Stage 1. The influent water was sampled from the settled water ducts at the plant.

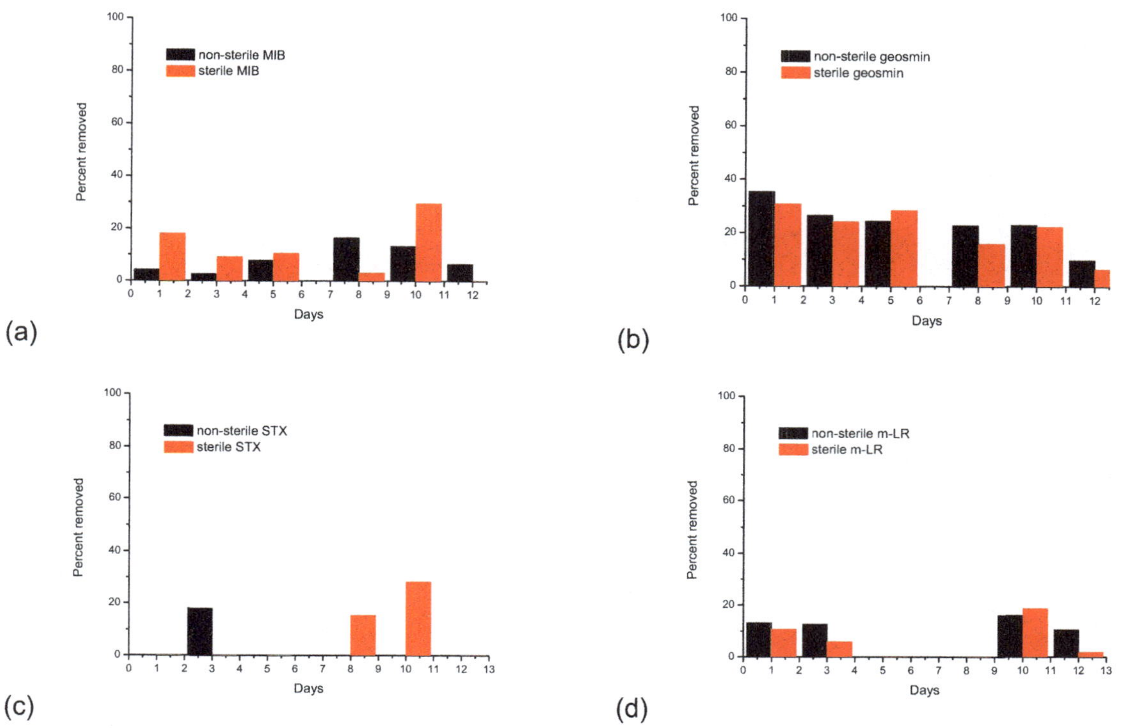

Figure 42: Column experiments, percent removals over 12 days: MIB, initial concentration 48 ng/L (a); geosmin, initial concentration 40 ng/L (b); saxitoxin equivalents, initial concentration 2.6 μg/L (c); microcystin LR, initial concentration 6.4 μg/L (d). Zero and negative results not shown.

The results suggest no measurable biological removal of these metabolites during laboratory column filtration.

3.2.2.2 Batch experiments

Settled water from Grahamstown WTP was spiked with MIB, geosmin, STXeq, mLR and mLA and cylindrospermopsin. Another water sample was sterilised prior to spiking. The concentrations of the metabolites sampled over 15 days are shown in Figure 43. Any decrease in concentration of the non-sterile sample indicates the presence of degrading organisms in the influent water to the filters at the full scale, and the potential of these organisms to colonise subsequent filtration media.

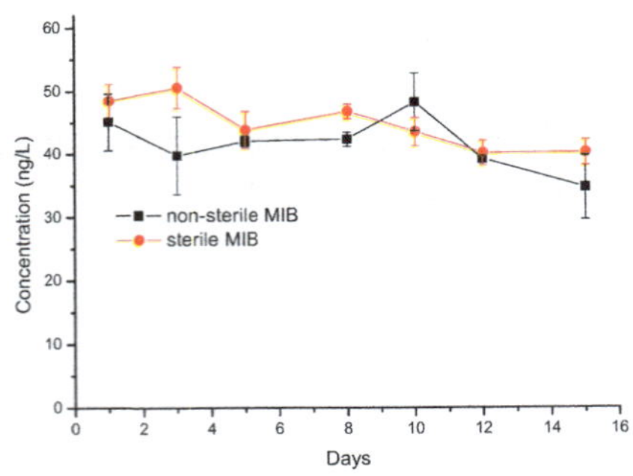

(a)

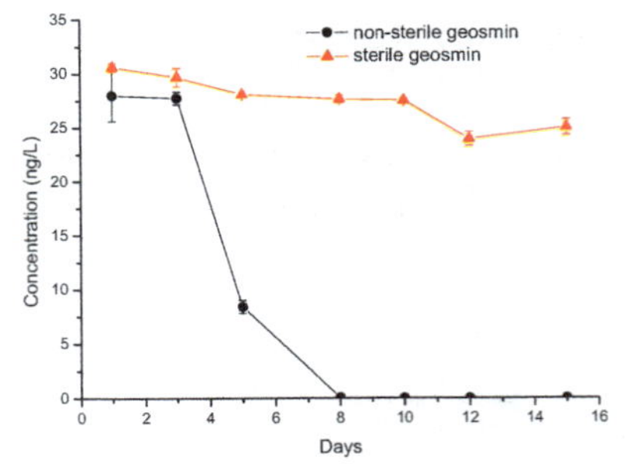

(b)

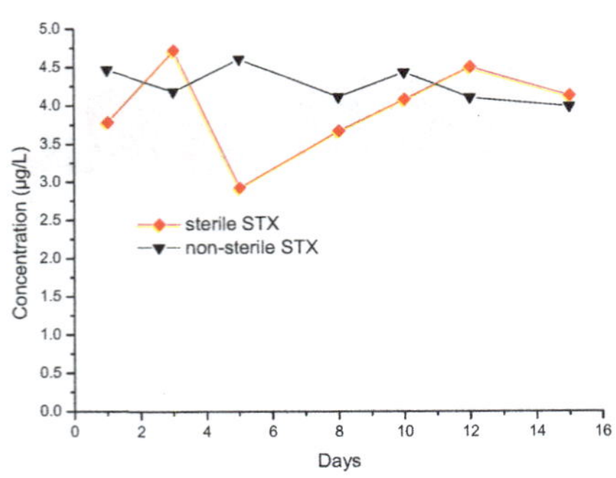

(c)

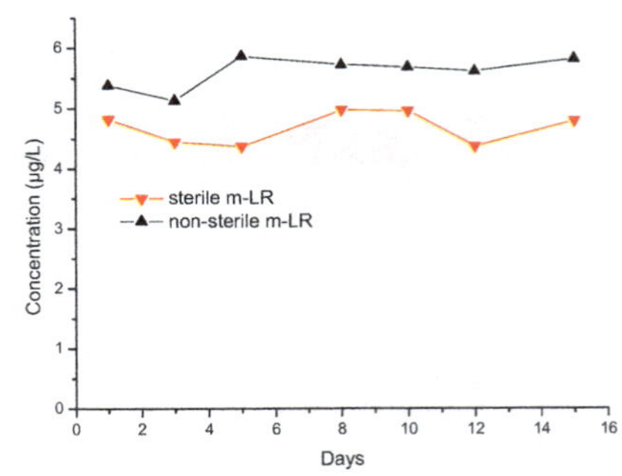

(d)

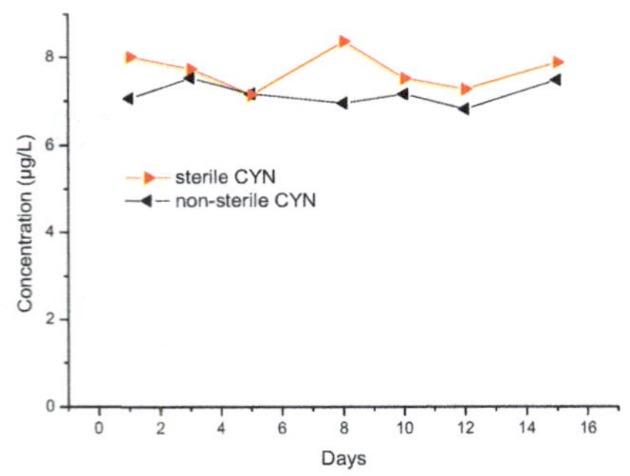

(e)

Figure 43: Batch experiments in settled water, concentration as a function of time: MIB (a); geosmin (b); STXeq (c); microcystin LR (d); cylindrospermopsin (e).

Relatively rapid degradation of geosmin occurred after a lag phase of one day in Grahamstown WTP settled water indicating the presence of degrading bacteria. This suggests that, given the suitable conditions it is possible that these bacteria might colonise the filters and provide biological removal. No degradation of the other metabolites occurred in the batch experiments, indicating the absence of degrading organisms in the influent to the filters at this WTP.

3.2.2.3 Summary

Although the filters are not currently functioning for the biological removal of the metabolites tested, geosmin-degrading bacteria are present in the influent to the filters suggesting that the filters at Grahamstown WTP have the potential to function as biofilters given suitable conditions for colonisation.

3.2.3 Microbiological analysis

Figure 44 shows the results of testing for copies of 16S rRNA, used as an indicator of biomass, on the filter media for the five media samples. Samples were taken prior to column testing and after the 12 day test for both sterile and non-sterile media. As DNA would still be present on the sterile samples, results for non-sterile and sterile samples could be expected to be similar, unless some of the dead biofilm had sloughed off the surface. In addition, we would not expect a large increase in biomass over the period of the testing, therefore the pre- and post- trial samples can be taken as duplicates to indicate the reproducibility of the procedure.

Figure 44 (a) gives the data on a log10 vertical axis to emphasise the difference between the negative controls and the low values of copies/g. Figure 44 (b) shows the data plotted with linear y-axis scale to display differences in the data.

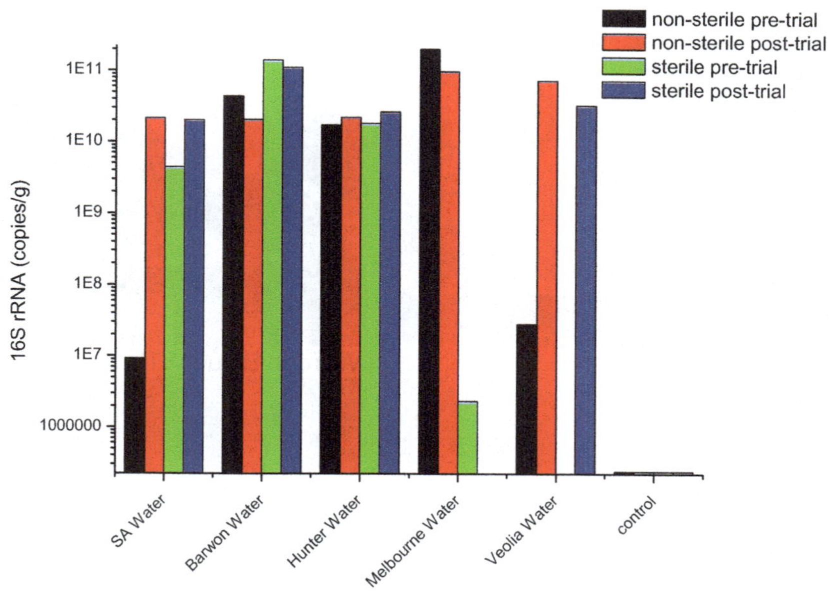

(a)

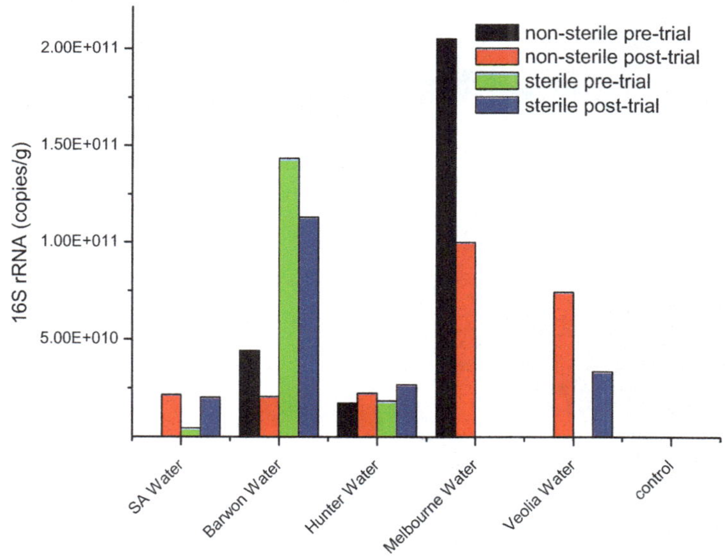

(b)

Figure 44: Copies of 16S rRNA per gram of media, prior to the column testing (non-sterile pre-trial) and after the 12 day test (non-sterile post-trial). Methods for analysis and controls are given in Hoefel et al. (2005).

As can be seen in Figure 44 (a), all samples contained 16S rRNA at concentrations significantly above the negative control, therefore all filter media contained some biological material. However, some of the samples showed much lower concentrations than would be expected compared with other samples from the same plant (e.g. SA Water and Veolia non-sterile pre-trial). Although the samples from the SA Water and Hunter Water filters appear to have a lower biomass measured by this technique, the large variation in results overall suggests that this difference may be within experimental error. No mlrA gene was detected, suggesting the presence of microcystin degraders in the media was minimal.

3.2.4 Comparison of media

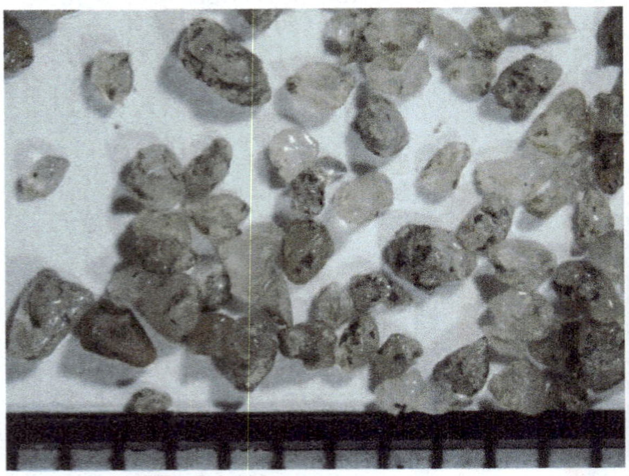

Morgan WTP, SA Water

Wurdee Boluc WTP, Barwon Water

Grahamstown WTP, Hunter Water

Winneke WTP, Melbourne Water

Illawarra WTP, Veolia Water

Figure 45: Filtration media. Bottom scale, 1 mm graduations.

One parameter that has been shown to have an effect on the biomass, and potentially the biological filtration potential, is the size of the filter media. Smaller particles have a larger surface area, and therefore can potentially sustain a higher biomass than the same volume of medium of a larger size. Figure 45 shows the media particles compared with a scale of 1 mm graduations. From these photos, the relative size of the sand particles is
Grahamstown WTP, Hunter Water > Winneke WTP, Melbourne Water > Morgan WTP, SA Water.
The anthracite particles from Illawarra WTP, Veolia Water and Wurdee Boluc WTP, Barwon Water appear to be of similar size and both are larger than the sand media samples.

3.2.5 Summary and conclusions

The BFP test was applied to the filters of five water treatment plants, the results are summarised below:

- All filtration media sustained some biomass, indicated by 16S rRNA analysis; however the variability in the results did not allow for direct comparison between the media
- No mlrA gene was detected in the DNA of the biomass, suggesting the presence of microcystin degraders in the media was minimal
- Morgan WTP sand was the only filter medium displaying biological removal of any of the metabolites investigated. Geosmin was removed 50-60% above the physical removal shown by a parallel sterile column filter. This suggests some lessening of the biofiltration efficiency of this sand compared to previous studies, as it had been shown to degrade both MIB and geosmin to below detection under similar conditions. This effective biological filtration was partly attributed to the small particle size of the sand used in the Morgan filters, which were shown to sustain a flourishing biomass (McDowall 2008). Batch testing revealed that geosmin, MIB and microcystin degrading bacteria were present in Murray River raw water for this plant but were not present in the influent water to the full-scale filters. It is possible that current operational conditions at the plant are preventing viable degrading bacteria reaching and re-seeding the filters, thus reducing the biological filtration efficiency. Investigations are continuing
- Grahamstown WTP settled water was the only filter influent water that contained bacteria capable of rapidly degrading geosmin. This result suggests that the filters have the potential to degrade geosmin if sufficient bacteria colonise the filters. However, the larger sand particle size used at Grahamstown WTP may prevent the development of the critical biomass required for effective biological filtration.
- Although the other filtration media contained some biomass, none displayed biological removal through filtration, and the influent water at the full scale did not contain degrading organisms; therefore, under the current conditions it is unlikely that biological degradation of metabolites would occur during rapid sand filtration at these plants.
- In the absence of disinfection all filter media will sustain a biofilm. The ability of the biofilm to remove cyanobacterial metabolites will depend on the number and composition of the colonising bacteria, which in turn depends on the entire biomass in the filter (surface area available for sustaining a biofilm) and the presence of degrading bacteria in the influent to the filter that can remain viable and thrive in a biofilm. From these tests it appears that the conditions required for the establishment and maintenance of a degrading biofilm may be atypical.

3.3 Investigation of potential impacts on the extent of the lag period for MIB and geosmin removal

3.3.1 Scenario 1 – Presence and absence of metabolites

Samples of anthracite and sand from Morgan WTP filters were collected in July 2013 for laboratory-scale column testing using Morgan settled water as the influent. Figure 46 shows the results of the column testing with continuous spiking of geosmin and MIB over 130 days. These results indicate that within 8 days both the sand and anthracite media removed greater than 50% of both taste and odour compounds. This indicates the lag period in this case was short and prolonged exposure to metabolites may decrease lag period and increase the removal. However, on day 43 the removal of both compounds was significantly impacted, especially in the anthracite media where removal was down to physical removal levels (Elhadi et al. 2004). Further investigation showed that there was no operational change at the WTP. The decrease in removal rates may suggest that there was a change in the bacteria present within the biofilm as operational conditions including water quality had not changed. Although the removal for the sand media did not drop below 50% for either MIB or geosmin, the removal observed for the anthracite media dropped below this threshold for both compounds. Continual running of each of the filter media columns showed recovery to greater than 70% removal. An increase in MIB removal from 15% to 74% for the anthracite media over approximately 90 days was observed, and an increase from approximately 55% to 91% for the sand. Geosmin removals also increased, from 39% to 85% for anthracite and 75% to 90% for the sand.

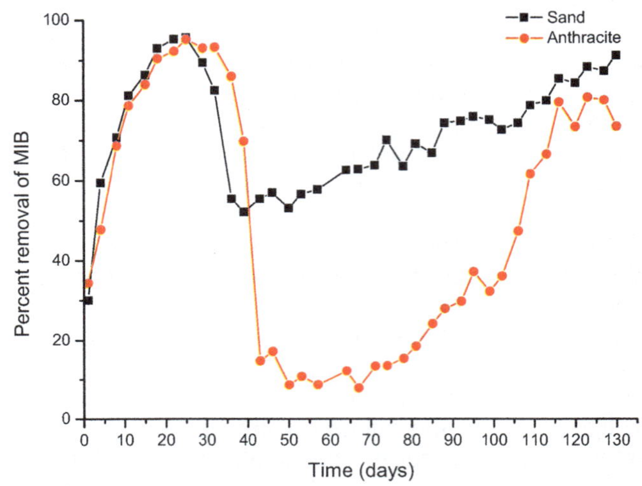

(a)

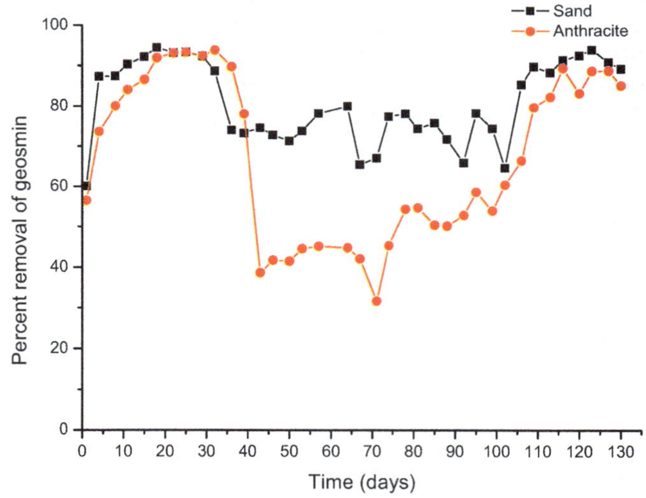

(b)

Figure 46: Laboratory column experiments, percent removals: MIB (a) and geosmin (b).

These results show higher biological degradation than was observed during the pilot trials (section 3.1.3). The laboratory-scale columns showed >40% removal of MIB and >70% removal of geosmin through both media after 4 days. The results in combination with pilot trials and observations at the full-scale plant suggest that the degrading biofilm may not be as robust as previously thought, and removals of MIB and geosmin may be significantly affected by unidentified changes in biofilm health, water quality or other operational conditions.

Following this trial period both filter media columns were backwashed (approximately 10-15 minutes) with chlorinated water (3 mg/L), to inactivate the bacteria present in the biofilm, until a residual (minimum 2 mg/L) was observed in the wash water (water that has passed through the filter media). Columns were then backwashed with deionised water, to remove any residual chlorine, and fed with settled water with continual presence of MIB and geosmin, to determine the lag period following inactivation of the biofilm. Figure 47 shows results of the column testing, post chlorination, over 36 days. The results indicate that the lag period for geosmin is short with greater than 50% removal achieved for both sand and anthracite within 4 days, while it took longer, 22 days, for the removal of MIB to reach 50%. This result suggests that the biofilm on both media could recover from a chlorination event within 4 weeks.

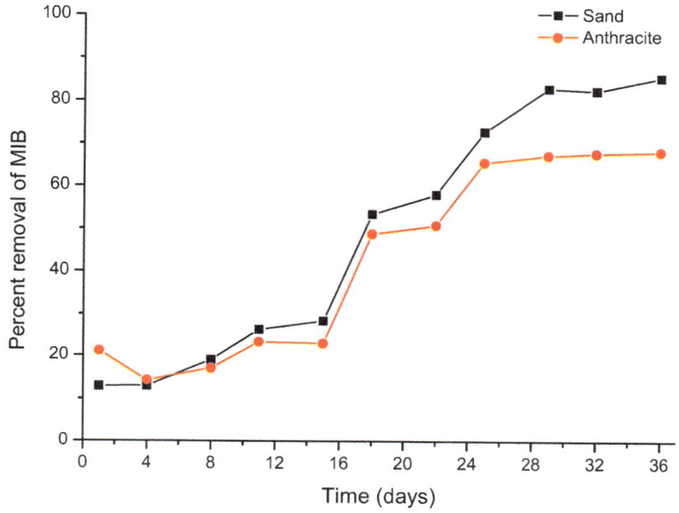

(a)

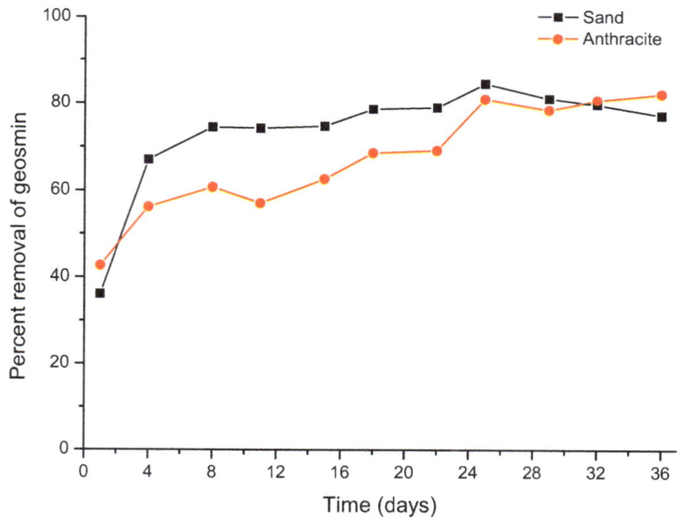

(b)

Figure 47: Laboratory column experiments post chlorination, percent removals: MIB (a) and geosmin (b).

Following the 36 day period, both filter media were again backwashed with chlorinated water, to inactivate the bacteria present and fed with settled water once again, however this time no MIB and geosmin were added. After 22 days, the lag period as determined from Figure 47, MIB and geosmin were added to the feed water to determine removals. The aim was to compare the lag period observed in the presence and absence of the metabolites after inactivation of the degrading biofilm. Figure 48 shows that when the metabolites are continuously present in the influent after biofilm inactivation the removals after 22 days are significantly higher than when the metabolites are not present. MIB removals were 30% for both the sand and anthracite media in the absence of metabolites while in the presence of metabolites the removal of MIB was 58% and 51% for sand and anthracite respectively. Geosmin removals were higher overall as expected (Elhadi et al. 2004) with 52% and 46% in the absence compared to 79% and 69% for sand and anthracite in the presence of the metabolite.

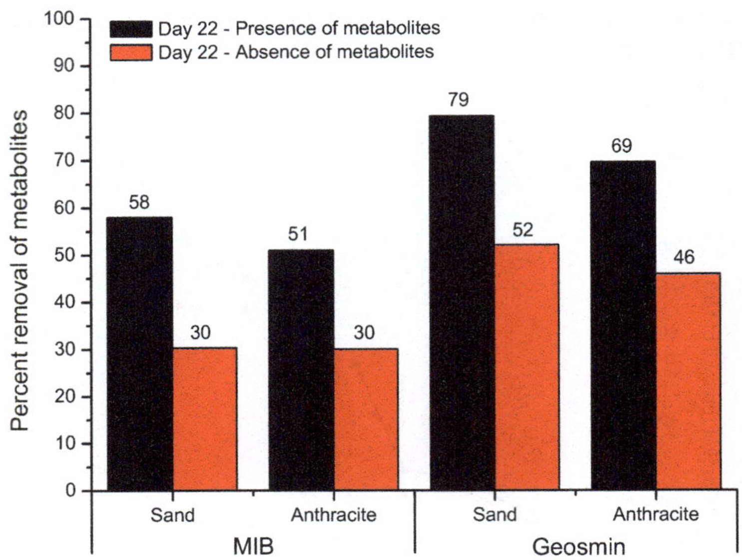

Figure 48: Comparison of day 22 results in the presence and absence of metabolites.

Batch experiments were completed using the settled water from Morgan WTP. No degradation of either MIB or geosmin occurred, indicating the absence of degrading organisms in the influent to the filters (Figure 48).

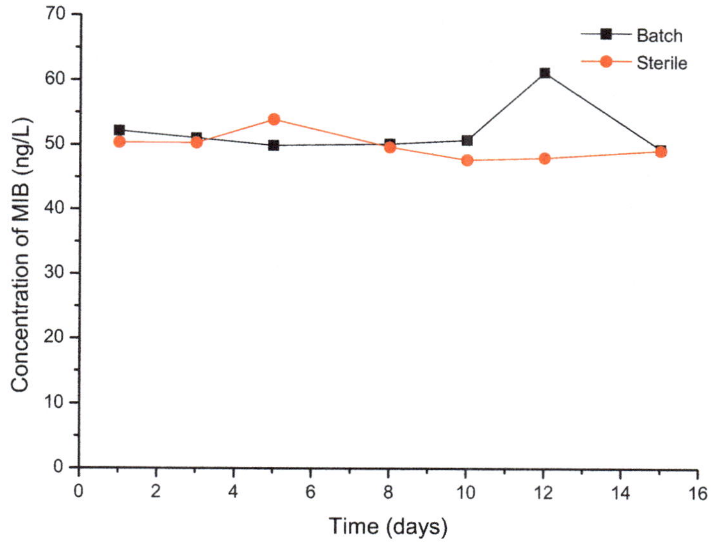

(a)

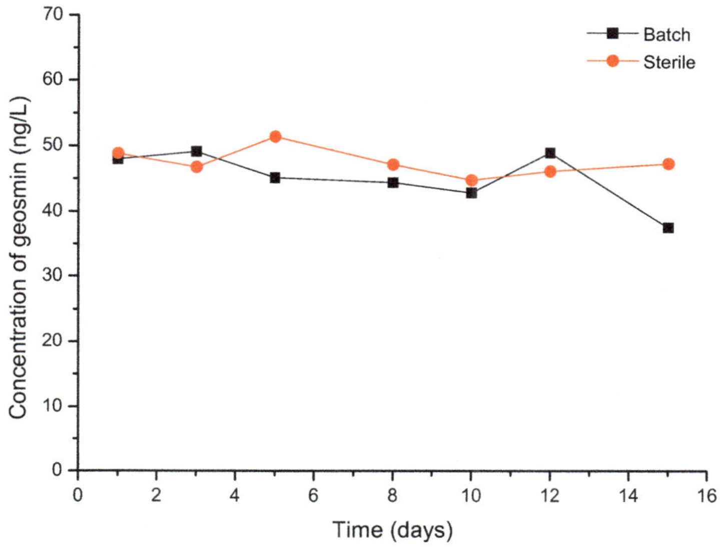

(b)

Figure 49: Batch experiments in settled water, concentration as a function of time: MIB (a) and geosmin (b).

Additional batch experiments were conducted in the raw water drawn from the Murray River (inlet to the WTP). Results, shown in Figure 50, are comparable with those found during the BFP test where organisms capable of degrading MIB and geosmin were present in the inlet water to this plant.

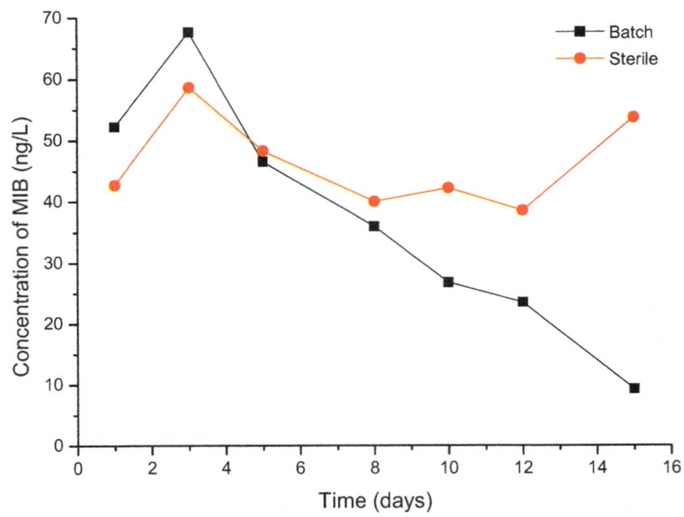

(a)

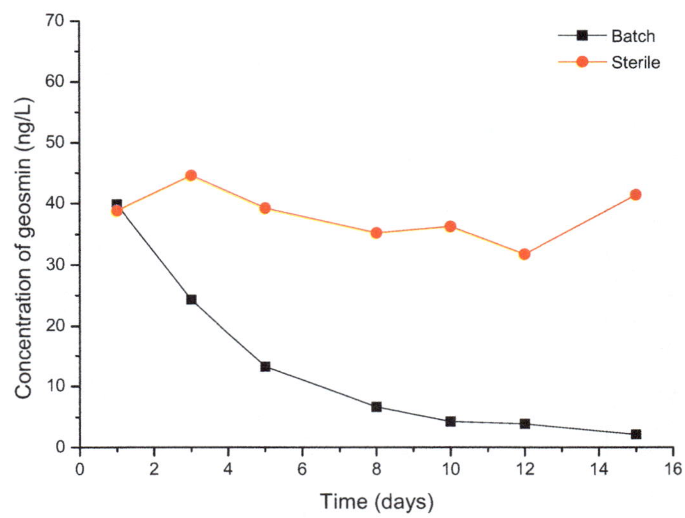

(b)

Figure 50: Batch experiments in raw water, concentration as a function of time: MIB (a) and geosmin (b).

The results from this scenario support the findings of the BFP test reported in section 3 where it was suggested that operational change at the WTP to improve coagulation and reduce turbidity also reduced the numbers of degrading bacteria in the settled water and therefore entering the filters. This change impacted the biofilm as degrading bacteria were reseeding the filters, further suggesting that ongoing seeding of degrading bacteria may be required for continual effective biological filtration of metabolites.

3.3.2 Scenario 2 – Seeding with indigenous bacteria

During the laboratory-scale column study performed at Happy Valley using settled water as the influent, no biological degradation of metabolites was observed over a 10 month period. A batch experiment was run on Happy Valley raw water to determine if MIB and geosmin-degrading organisms were present in the inlet to the plant. Results showed efficient degradation of geosmin to below the limit of reporting over a 16 day period and just over 50% degradation of MIB (Figure 51). Two of the media columns were chosen from the initial filter media study for this work (sand 6.5 and filter coal 0.8-1.0) with the aim to seed with indigenous bacteria directly using Happy Valley raw water and improve biological removal. Filter media details are shown in Table 5.

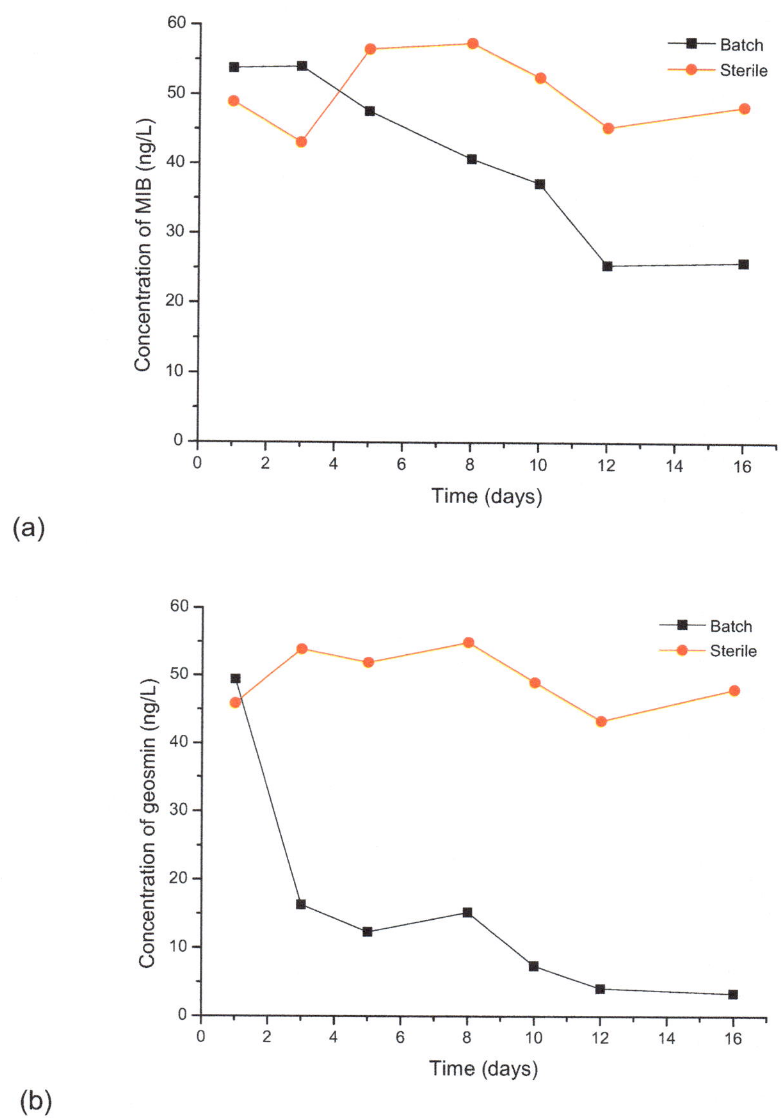

Figure 51: Batch experiments in Happy Valley raw water, concentration as a function of time: MIB (a) and geosmin (b).

Table 5: Details of filter media

Material	**Supplier, product**	**Particle size (mm)**
Sand	Riversands, 6.5	0.85-0.95
Bituminous coal	Riversands, filter coal	0.8-1.0

The media were chosen as neither displayed biological removal of the metabolites, and the particle size is commonly used in full-scale filters.

The columns were fed for approximately three months on Happy Valley raw water to allow biofilm development in the absence of metabolites, followed by an additional 12 months with MIB and geosmin intermittently spiked during this period. MIB and geosmin were then spiked continuously into the reservoir feed water at a target concentration of 100 ng/L of each. Samples were collected from the influent and effluent of each column twice a week for MIB and geosmin analysis over a period of 50 days. The percent removals are shown in Figure 52. Removals of both compounds was initially greater than 90% in both filter media, however the removal decreased in the sand column for MIB after a week, while the removal for filter coal remained above 90%. During the same period the removal of geosmin through the sand filter decreased to below 70%, then recovered within three weeks to the same removal as the filter coal. Similarly to scenario 1, no cause for this decrease in biodegradation could be ascertained.

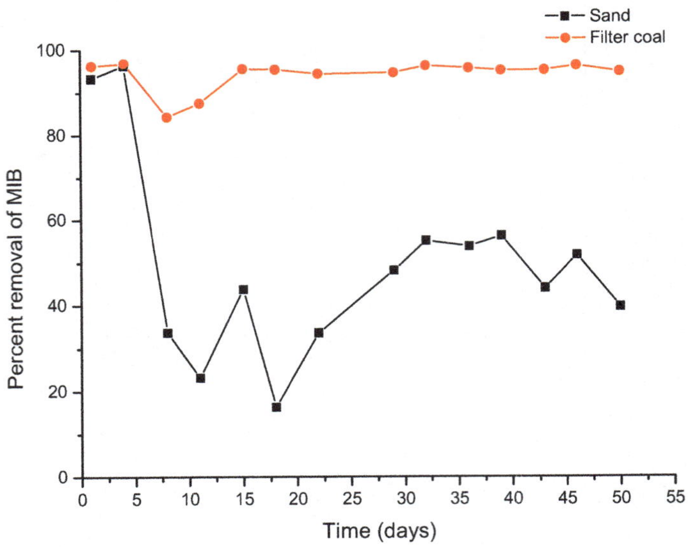

(a)

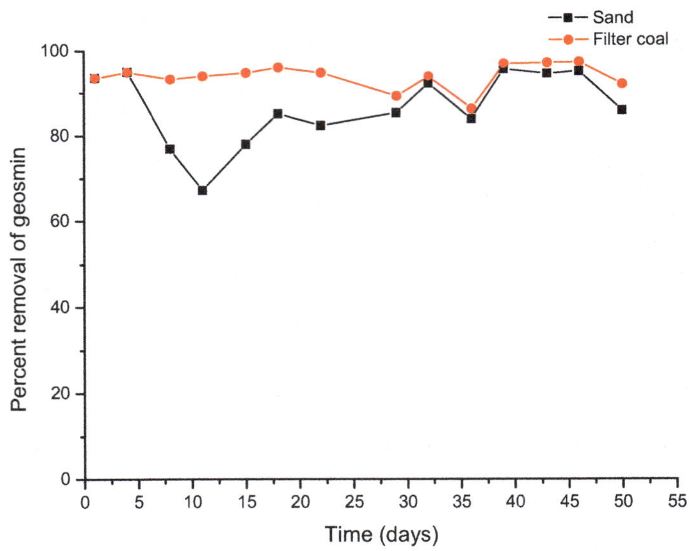

(b)

Figure 52: Laboratory column experiments for Happy Valley raw, percent removals: MIB (a) and geosmin (b).

3.3.3 Scenario 3 – Presence and absence of degrading bacteria on filter media

Following scenario 2, Happy Valley raw water was autoclaved (121°C) to inactivate degrading bacteria present in the influent water. The columns were fed with the sterile raw water for four weeks and then spiked with MIB and geosmin to determine the effect of the absence of degrading bacteria in the feed water. The reasoning was that this might occur in the full-scale if optimised coagulation was applied resulting in an increase in removal of micro-organisms.

The results show 95% removal of geosmin during most of the spiking period, with a small decrease in removal at day 21 (Figure 53 a). The initial removal of MIB was 71% for sand and 96% filter coal – suggestive of biological degradation (Figure 53 b). In contrast to the removal of geosmin, the removal of MIB decreased markedly during the spiking period and on day 21 the effluent contained over nine times the concentration of MIB in the influent. It is possible that the absence of degrading bacteria in the influent for over four weeks caused a change in bacterial population in the biofilm to a situation where producers of the compounds were established.

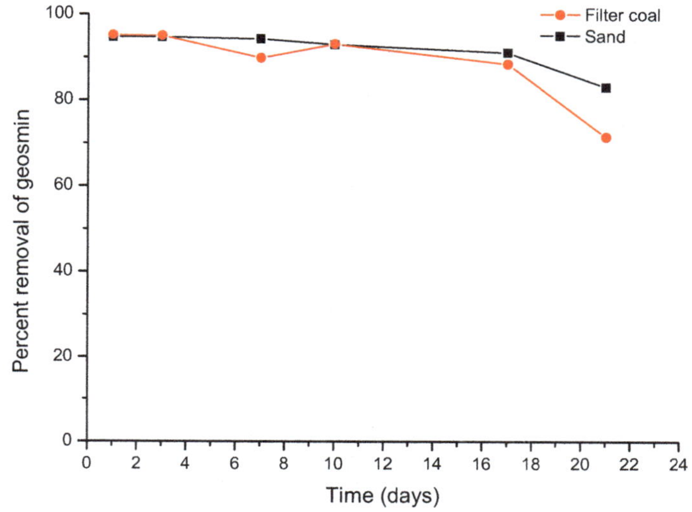

(a)

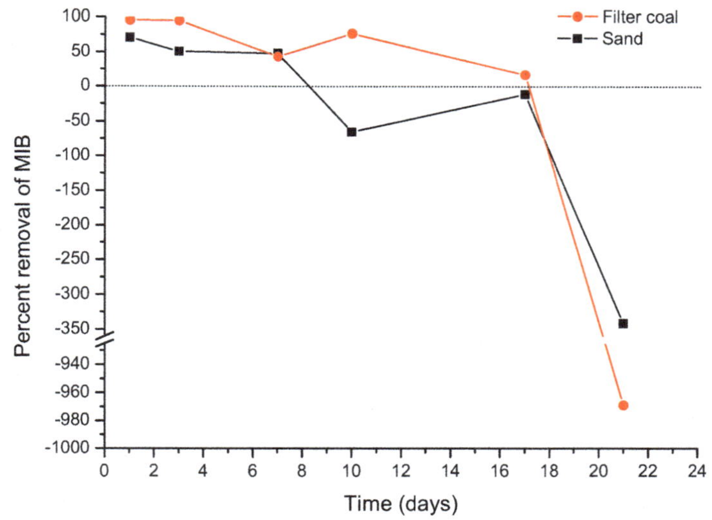

(b)

Figure 53: Laboratory column experiments for sterile Happy Valley raw, percent removals: geosmin (a) and MIB (b).

After the spiking period, samples were taken during backwashing of the laboratory columns for MIB and geosmin analysis, with results shown in Table 6. The feed water for the columns was switched back to Happy Valley non-sterile raw water. Samples were taken regularly from the influent and effluent as well as during backwashing. Analysis of backwashing samples confirmed the theory of MIB producers present in the media, as concentrations in the backwash water from both the sand and filter coal columns were higher than influent levels. Effluent sample concentrations were also found to be higher than influent concentrations, and the concentrations had not diminished indicating that it is not released from the media (desorption), (Table 7).

Table 6: MIB and geosmin results for laboratory column backwashing (20-Sept-2013)

Sample	MIB (ng/L)	Geosmin (ng/L)
1 min BW Sand	1380*	42
5 min BW Sand	11500*	320*
1 min BW Filter coal	270*	17.7
5 min BW Filter coal	65	<4.0

*Above calibration of analysis
Backwash (BW)

Table 7: MIB and geosmin results for laboratory column sampling (11-Oct-2013)

Sample	MIB (ng/L)	Geosmin (ng/L)
Influent	8	5
Sand Filtrate	133	9
Filter coal Filtrate	146	8
1 min BW Sand	920 *	25
5 min BW Sand	124	<4
1 min BW Filter coal	250 *	6
5 min BW Filter coal	49	<4

*Above calibration of analysis
Backwash (BW)

Samples of the biofilm present within the feed reservoir were analysed and it was found that benthic cyanobacteria were present. They were identified using phase contrast microscopy (1000x magnification) as *Lyngbya spp.*, a genus that forms long unbranching filaments inside a rigid sheath. Recent work has shown that at least one strain of *Lyngbya* has the ability to produce geosmin (Schrader and Blevins 1993), however at this stage it is unknown if the strain found within the reservoir is a producer. As part of another complementary project, also supported by WaterRA (ARC Linkage project number LP120200587) "Bad tastes, odours and toxins in our drinking water reservoirs: are benthic cyanobacteria the culprits?" the strain will be isolated to determine if this strain is a producer of MIB or geosmin, or both, or neither.

In summary, the results suggest that the change in water quality, i.e. absence of degrading bacteria, caused a shift in bacterial population or bacterial equilibrium from degraders to producers. Initially degraders were present for both taste and odour compounds. However, following the change to autoclaved feed water MIB concentrations were higher in the effluent than the influent, suggesting producers of the compounds were dominant within the biofilm. Continued analysis during backwashing of media indicated that the increase in MIB in the effluent was not caused by release from the media (desorption).

This effect has been noted previously in both laboratory and full-scale experiments. High concentrations of geosmin were found in backwash water from the Morgan WTP filters, and the genes responsible for the production of the compound were also identified (2012, unpublished data). In addition, changing the influent water source from Morgan water to Happy Valley settled water during a laboratory-scale experiment using Morgan filter sand caused a significant increase in the geosmin concentration during backwash. The experiment was abandoned as the column continued to produce geosmin (McDowall 2008, unpublished data). It is possible that variable removals of T&O at the pilot and full scale could be caused by the simultaneous production and removal of metabolites.

Following the finding that MIB producing organisms were present within the filter media biofilm, both filter media columns were backwashed with chlorinated water (3mg/L), to inactivate the bacteria until a residual was observed in the wash water. It was understood that chlorination would affect the producers as well as any degraders present. This could be a management strategy employed at a full-scale treatment plant if a similar situation arose. Columns were then backwashed with deionised water to remove any residual chlorine and fed with raw water spiked with MIB and geosmin. Figure 54 shows the results of the column testing over 29 days after chlorination.

The lag period for geosmin was four days, based on the definition of lag within this project, with greater than 50% removal achieved for both sand and filter coal. It took longer for the removal of MIB to reach 50%, 15 days for filter coal and 22 days for sand. These result showed that chlorination of the media and subsequent exposure to the metabolites resulted in the control of the MIB producers and the re-establishment of an effectively degrading biofilm.

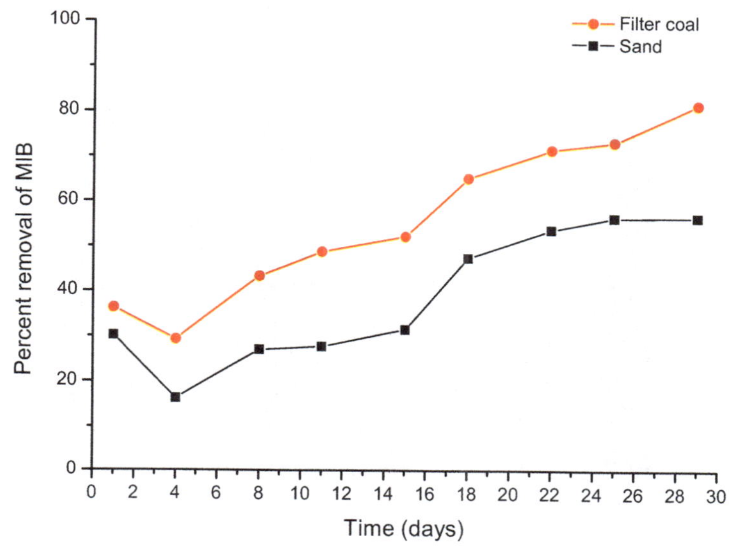

(a)

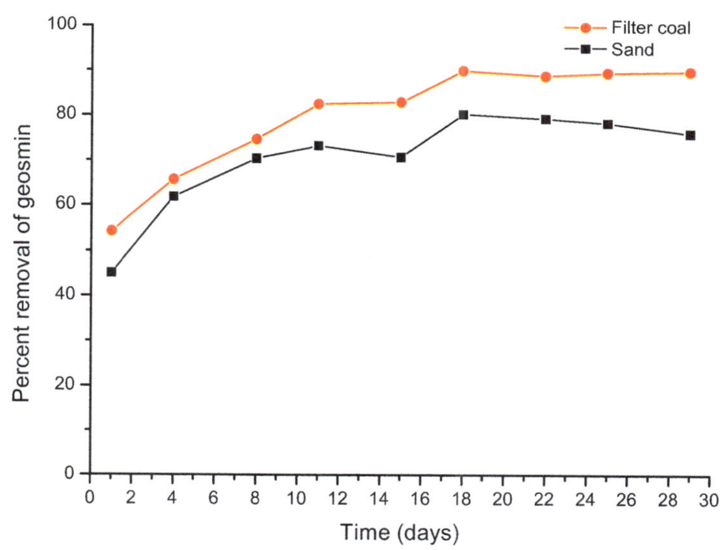

(b)

Figure 54: Laboratory column experiments post chlorination, percent removals: MIB (a) and geosmin (b).

Following the 29 day trial, both filter media were backwashed with chlorinated water, to again inactivate the bacteria, and the columns were fed with raw water with no MIB and geosmin added. After 22 days, MIB and geosmin were added to the feed water to determine the difference in removal in the absence of metabolites. Comparison of the results on day 22 showed that when the metabolites were continuously present the removal rates after 22 days were higher than when the metabolites are not present (Figure 55). MIB removals were 34% for sand and 42% for filter coal in the absence of metabolites with higher removals of 54% and 72% in the presence of metabolites. Geosmin results were higher than MIB overall with 55% and 62% for sand and filter coal in the absence and 79% and 89% for sand and filter coal respectively in the presence of metabolites.

These results are similar to those found in scenario 1, further suggesting that systems where filtration media experience more regular episodes of taste and odour issues are more likely to have higher removals than a plant which has not experienced a taste and odour issue in the past.

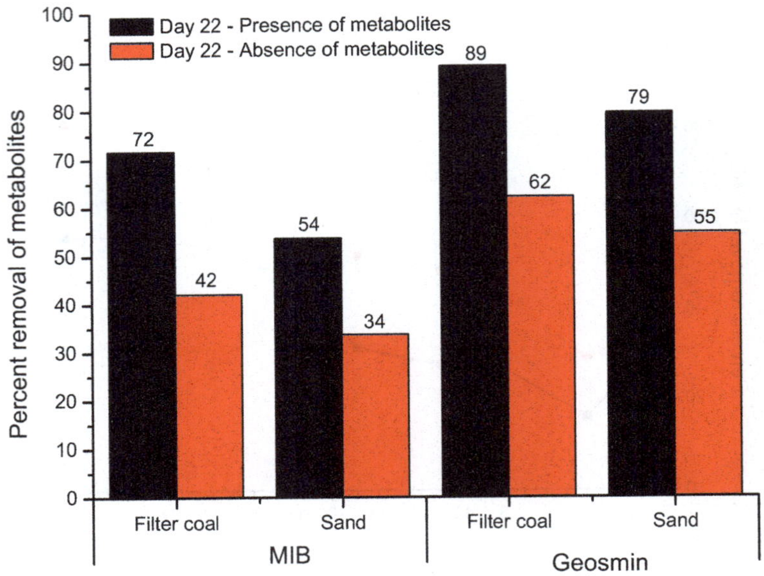

Figure 55: Comparison of day 22 results in the presence and absence of metabolites.

3.3.4 Summary

These scenarios have shown that it is possible to encourage biological degradation by acclimation (scenario 1) and seeding (scenario 2); however the conditions required for the establishment and maintenance of a degrading biofilm may not always be maintained during changes in operational conditions. For example, if the indigenous bacteria are not present on a continuous basis, under reasonably constant conditions, the biological balance within the biofilm may change and result in deleterious effects on water quality, including breakthrough of taste and odour compounds (scenario 3).

Overall, in the absence of disinfection, all filter media will sustain a biofilm; however the ability of the biofilm to remove the cyanobacterial metabolites, MIB and geosmin, will depend on the number and composition of the colonising bacteria, which in turn depends on the entire biomass in the filter (surface area available for sustaining a biofilm) and the presence of degrading bacteria in the influent to the filter that can remain viable and thrive in a biofilm.

4 SUMMARY AND CONCLUSIONS

The objective of this project was to develop operational guidelines for the implementation of biological filtration for the removal of cyanobacterial metabolites at the full scale.

The research approach was designed to address five hypotheses:

- All filters have the potential to function as biological filters, given the right conditions (Sections 4.1.1, 4.1.2, below)
- Media surface character and particle size will influence biodegradation (Section 4.1.1)
- The existing biological potential at the full scale can be measured at the laboratory scale (Section 4.1.2)
- Operations can be modified to allow biological filtration to occur (Section 4.1.1, 4.1.3)
- The lag period can be reduced by seeding with degrading bacteria (Section 4.1.3)

The project hypotheses were tested using the approach described below.

4.1 Summary of Outcomes

4.1.1 Identification of optimum operating conditions to enable effective biological filtration

Previous research has shown that a range of filter media, of different composition and size, will support a healthy biofilm in the absence of a disinfectant residual. Various studies suggested that topography and size of the media influence formation of a biofilm, with smaller particle and rougher surface facilitating the development of a higher biomass.

The aim of this study was to compare different particle sizes of the same medium, and the same particle size of different media to determine the effect of filtration media particle size and surface characteristics on the implementation of biological filtration. This effect had not been studied systematically previously. In laboratory-scale columns we investigated sand, filter coal and granular activated carbon (GAC), of two particle sizes each, and one other substrate, Filtralite, which is promoted as a specialised filtration medium made from expanded clay particles. The expanded clay particles offered a novel surface chemistry, with pores of a size that should also be amenable to biofilm formation.

The laboratory columns were fed with settled water from the Happy Valley WTP for 10 months. During this time the columns were backwashed on a weekly basis and spiked periodically with cyanobacterial metabolites MIB, geosmin, saxitoxins, microcystin LR and LA (mLR and mLA) and cylindrospermopsin. Although all media developed a healthy and diverse biomass, as shown by scanning electron microscopy (Figures 29-34), the removals of metabolites and NOM were at levels representative of physical processes, with no evidence of biological degradation (Figures 12-27). Under these conditions the GAC was effective for MIB and geosmin removal, with average 98 and 100% removal respectively over the 10 month study. In agreement with previous findings, the removal of toxins was high at the time of commissioning (100% for microcystins and cylindrospermopsin and 85% for STXeq measured by ELISA) then decreased over the 10 month trial. The reduction in adsorption for the toxins was in the order mLA>mLR>cylindrospermopsin>STXeq. At the completion of the trial removal of mLA by both GACs was of the order of 20%, only slightly higher than the physical removals displayed by the non-porous media.

In previous pilot trials using Happy Valley settled water biological removal of MIB and geosmin was observed on sand after 10 months (Figure 3). After 10 months this comprehensive trial was terminated with the conclusion that the different particle sizes and surfaces tested did not influence the development of a biofilm capable of degrading any of the metabolites over this period. While the removals of MIB and geosmin observed on the GAC filters at the conclusion of this study may have been influenced by biological as well as physical processes, the removal of the toxins was through physical processes alone.

Two of the media were retained for investigation of the lag period prior to the commencement of effective biological filtration (Section 4.3).

Pilot plant investigations were undertaken at the Morgan WTP using sand from the full-scale filters. From previous column work it was assumed that the filters were functioning as biological filters and would remove both MIB and geosmin to below detection. This investigation was designed to determine the effect of hydraulic loading on the removal of the MIB and geosmin in a full-scale biological sand filter.

At filtration rates equivalent to normal full-scale operating conditions removals of both T&O were consistent with physical removal only, no additional removal attributable to biological processes was observed. This was unexpected as the plant had continued to remove geosmin effectively during recent challenges. However, during several episodes of high MIB levels entering the plant (up to 63 ng/L), only 40-60% MIB removal was observed at the full scale, some of which could be attributed to powdered activated carbon application. At lower filtration rates removals of geosmin increased with increasing contact time. These results were consistent with biological removal, although the observed removals were lower than expected from previous work. The differences observed between the full-scale removal of geosmin and at the pilot scale may be a result of the media. In the pilot columns only sand was used whereas the Morgan filters are sand/anthracite. It is possible that higher biological removals are experienced within the biofilm on the anthracite and this effect would not have been observed in this study.

4.1.2 Development of the "Biological Filtration Potential Test"

Effective biological removal of MIB and geosmin had been observed at Morgan WTP for over 14 years (see Figure 4). The only previous interruption to this removal occurred during a period where monochloramine was present in the backwash water and the biofilm was damaged (Figure 4 end 2005-end 2006). Once the process was reversed the biofilm re-established within a few months and geosmin was again effectively removed (Figure 4, 2006-2014). This effect was also seen at the full scale for MIB, although the removals were lower than for geosmin (data not shown). Research over the past 10 years has shown conclusively that the media filters (sand and anthracite) at Morgan WTP function as biological filters for MIB and geosmin, and it was hypothesised that other rapid filters could function in a similar manner under the appropriate conditions. However, at the full scale it is difficult to determine what processes in the plant are contributing to the removal of metabolites unless systematic investigative sampling is undertaken during a T&O or cyanotoxin event. A laboratory-scale procedure was developed to determine the extent of biological degradation occurring in existing filters in treatment plants managed by partner organisations, and the potential for a biofilm to develop that may be capable of degrading the metabolites. Five treatment plants were tested using the procedure, which included the metabolites MIB, geosmin, mLR and LA, saxitoxins and cylindrospermopsin. Only Morgan WTP sand showed evidence of biological filtration and only for geosmin. The results also showed that at the time of sampling there were insufficient degrading bacteria in the influent to the filters at Morgan WTP to degrade the compounds in a batch experiment undertaken over two weeks. A similar batch experiment undertaken on the raw water to the plant showed the presence of MIB and geosmin degraders. At Grahamstown WTP (Hunter Water) the influent to the filters contained geosmin degrading bacteria; however the filter media did not display additional removal due to biodegradation. The samples from the other three plants showed no evidence of degrading bacteria in the influent water to the filters or in the filter media itself.

4.1.3 Investigation of potential impacts on the extent of the lag period for MIB and geosmin removal

Two of the filter media tested in the experiments described in Section 4.1.1 were used for further investigations. In addition, samples of anthracite and sand from the full-scale filters at Morgan WTP were tested in laboratory-scale columns. The aim was to identify methods to enhance the effective removal of MIB and geosmin through biological filtration, and operational factors that could negatively impact the removals.

Three scenarios that may occur at the full scale were investigated:

4.1.3.1 Scenario 1 – Presence and absence of metabolites

It has been suggested in the literature that the rate of development of a biofilm capable of degrading metabolites (or the extent of the lag period) is dependent on the presence or absence of the metabolites themselves. In this work, samples of anthracite and sand from Morgan WTP filters were used for laboratory-scale column testing using Morgan settled water spiked with geosmin and MIB as the influent. The initial testing took place over 130 days. The results indicate that within 8 days both the sand and anthracite media removed greater than 50% of both taste and odour compounds. This indicates the lag period was short and prolonged exposure to metabolites can increase the removal. However, on day 43 the removal of both compounds was significantly impacted, especially in the anthracite media where removal decreased to physical removal levels. Continued running of each of the filter media columns showed an increase in MIB removal from 15% to 74% for the anthracite media and 55% to 91% for the sand over approximately 90 days. Geosmin removals also increased, from 39% to 85% for anthracite and 75% to 90% for the sand. The cause for the large impact on the biofilm was unknown as the water quality parameters and operating conditions remained the same throughout the trial.

After 130 days, when the degrading biofilm had re-established, the columns were chlorinated to deactivate the biofilm, and the spiking was resumed at the previous levels. The observed lag period for geosmin was short, with greater than 50% removal achieved for both sand and anthracite within a week, while it took longer (22 days) for the removal of MIB to reach 50%. The results suggested that the biofilm on both media could recover from a chlorination event within 4 weeks, similar to results seen at the full scale in 2005-2006. After 36 days the filters were again de-activated using chlorine and the trial was re-commenced in the absence of the metabolites. After a further 22 days MIB and geosmin were added to the feed water to determine removals. The aim was to compare the lag period observed in the presence and absence of the metabolites after inactivation of the degrading biofilm. The removals of both MIB and geosmin at 22 days after biofilm deactivation were lower when the biofilm was re-established in the absence of the metabolites. For both compounds the recovery of the biofilm appeared to be more rapid on the sand.

4.1.3.2 Scenario 2 – Seeding with indigenous bacteria

Although artificially seeding laboratory filters with cultured metabolite-degrading bacteria has shown some promising results, further laboratory-, pilot- and full-scale investigations have suggested that such advantages may be short-lived. As a result, testing was undertaken to determine whether artificially seeding with indigenous bacteria, with higher numbers than would normally be found in the influent water (raw water compared to settled water), would have a positive effect on the degradation of metabolites. In addition, as shown in the previous section, the presence of MIB and geosmin on a continuous basis may promote the increase in the numbers of degrading bacteria in the biofilm and thus improve the degradation efficiency of biological filters.

Two media columns were chosen from the initial filter media study for this work (sand 6.5 and filter coal 0.8-1.0) with the aim to seed with indigenous bacteria directly using Happy Valley raw water and improve biological removal. Batch experiments showed the existence of both MIB and geosmin degraders in Happy Valley Reservoir water.

The columns were fed for approximately 3 months on Happy Valley raw water to allow biofilm development in the absence of metabolites, followed by an additional 12 months with MIB and geosmin intermittently spiked during this period. MIB and geosmin were then spiked continuously into the reservoir feed water at a target concentration of 100 ng/L over a period of 50 days. Removal of both compounds was initially greater than 90% in both filter media, however the removal decreased in the sand column for MIB after a week, while the removal for filter coal remained above 90%. During the same period the removal of geosmin through the sand filter decreased to below 70%, then recovered within three weeks to the same removal as the filter coal. Similar to scenario 1, no cause for this decrease in biodegradation could be ascertained. The results showed that the indigenous degrading bacteria had established within the biofilm and effective biological filtration of MIB and geosmin occurred initially. In addition, after damage to the biofilm of an unknown cause, the biofilm recovered over a period of approximately one month. In this scenario the biofilm on the filter coal appeared to be less susceptible to damage than the sand biofilm as both were subject to the same conditions.

4.1.3.3 Scenario 3 – Presence and absence of degrading bacteria

The results obtained in the project to date, and full-scale data, suggested that changing water quality, in particular the number and type of bacteria entering the filter, could have a significant impact on the potential for the filter to achieve effective biological filtration. In this section of work the two columns used in the previous section were changed from Happy Valley raw water to sterilised raw water. Therefore the physical water quality remained the same (nutrients, NOM pH etc) but there were no active bacteria present in the influent to replenish the biofilm. The columns were fed with sterile raw water for 4 weeks and then spiked with MIB and geosmin for 3 weeks to determine the effect of the absence of degrading bacteria in the feed water.

The results showed 95% removal of geosmin during most of the spiking period, with a small decrease in removal at day 21. The initial removal of MIB was 71% for sand and 96% filter coal – also indicative of biological degradation. In contrast to geosmin, the removal of MIB decreased markedly during the spiking period and on day 21 the effluent contained over 9 times the concentration of MIB present in the influent. The results suggest that the absence of degrading bacteria in the influent for 4 weeks caused a change in bacterial population in the biofilm to a situation where producers of the compounds were established. After the spiking period the filters were fed with non-sterile Happy Valley Reservoir water for an additional 4 weeks in an effort to re-establish MIB degrading bacteria within the biofilm. During this period high concentrations of MIB were evident in the backwash water and the filter effluent of both media, while geosmin concentrations remained at background levels. Both media were then chlorinated to deactivate all bacteria, and the trials were resumed on non-sterile Happy Valley water spiked with MIB and geosmin. The removals increased over time for both metabolites, with the lag period – as defined by 50% metabolite removal – found to be 22 days for MIB on the sand filter. After 29 days the filters were chlorinated to deactivate the bacteria and the columns were fed with non-sterile Happy Valley Reservoir water in the absence of metabolites. As observed with the Morgan columns the time taken to reach >50% removal of T&O was longer in the absence of the metabolites. In this scenario both media showed similar trends although the percent removals on the filter coal were slightly higher.

4.2 Conclusions and recommendations

The establishment of a biofilm on filter media takes place within several weeks of commissioning. However, the establishment of a biofilm capable of degrading cyanobacterial metabolites to levels acceptable for consumers may take considerably longer.

The long-term removal of cyanobacterial metabolites in biologically active filters is dependent on a range of complex and interrelated factors including:

- The number of degrading bacteria present in the influent water
- The ability of those bacteria to colonise filtration media
- A critical mass of degrading bacteria being retained in the media
- The maintenance of water quality conditions conducive to a healthy, degrading biofilm
- The frequent/regular presence of the metabolites in the influent to the filters

Some of these factors can be controlled at the treatment plant, some are dependent on the raw water quality. In addition, factors that may be influenced by plant operations could conflict with other water quality goals. For example, optimisation of coagulation for improved particle removal may result in reduced bacteria numbers in the influent to the filters, and consequent decrease in the efficacy of biological filtration. In addition, results at the laboratory and full scale suggest that biological removal may be impacted rapidly by unknown factors, resulting in unforseen breakthrough into the treated water.

As a result, it is not possible to develop specific guidelines for the confident application of biological filtration for the removal of cyanobacterial metabolites at all treatment plants. However general recommendations that may increase the probability of successful biological filtration are given below, and suggestions are made to aid in the determination of the potential for the development of the process at a particular plant.

4.2.1 Can your filters achieve successful biological filtration?

If you answer "yes" to all of statements below there is the potential for the formation of a degrading biofilm on your filters:

- Our filtration medium has an average particle size < 1 mm
- There is no pre-chlorination or pre-filter chlorination at the WTP
- Metabolite(s) are regularly present in the raw water
- Metabolite(s) are regularly present at the inlet to the filters
- There are degrading organisms in the raw water
- Degrading organisms regularly reach the inlet to the filters

4.2.2 How can you improve the chances of biological filtration?

- Use smaller media particle size
- Cease pre-chlorination and/or pre-filtration chlorination if usually practiced
- Allow degrading bacteria to colonise filtration media – seed filters with indigenous bacteria if feasible
- If a degrading biofilm has established it will regenerate after severe damage comparatively quickly once suitable conditions for regrowth have been re-established

Granular activated carbon is the recommended medium for the implementation of biological filtration for the removal of cyanobacterial metabolites. The adsorbent allows for removal of metabolites during the often prolonged period prior to onset of biodegradation. However, the same conditions apply as above (4.2.1) – in the absence of these conditions adsorption of cyanotoxins will decrease over time, due mainly to continuous adsorption of NOM. The GAC may be rendered ineffective by "pre-loading" of NOM over a period of 12 months or less.

Table 8: Findings at a glance

Factor	**Operational Implication**
Particle Size	• Reducing particle size will increase the surface area available for biofilm growth
Acclimation	• WTPs that experience more regular episodes of taste and odours are more likely to achieve biological removal of the metabolites • Once a degrading biofilm is established operational changes may impact efficacy • Re-establishment of a biological removal on a damaged biofilm will occur more rapidly than on new filtration media
Water Quality	• Enhanced coagulation will remove up to 1 log of bacteria from feed water of filters → increasing time to develop a biomass capable of biological degradation • Reducing bacterial numbers in filter influent changes biodiversity of biofilm and may result in production of T&O Disinfection • Pre-chlorination will inactivate bacteria reducing chance of biological degradation • Pre-filter chlorination (used to increase filter run times) will sterilise the filters and prevent effective biological removal of metabolites

5 PUBLICATIONS AND PRESENTATIONS ARISING FROM THIS PROJECT*

5.1 Peer reviewed journals

- Ho L, Sawade E, and Newcombe G (2012) Biological treatment options for cyanobacteria metabolite removal – A review, *Water Research*, 46 (5), 1536-1548.

5.2 Conferences presentations/proceedings

- Sawade E, Ho L and Newcombe G (2014) Effect of water quality changes on biological filtration efficacy, In Proceedings. *4th National Cyanobacteria Workshop*, 22-24 September 2014, Adelaide, Australia.
- Sawade E, Vache M, Ho L, Tang T and Newcombe G (2012) Implementation of biological filtration for the treatment of cyanobacterial metabolites, In Proceedings. *American Water Works Association Water Quality Technology Conference (AWWA WQTC),* 4-7 November 2012, Toronto, Canada.
- Sawade E, Ho L, Hoefel D, McDowall B and Newcombe G (2012) Development and validation of the biological filtration potential test for the removal of cyanobacterial metabolites, In Proceedings. *AWA OzWater Convention and Exhibition*, 8-10 May 2012, Sydney, Australia.
- Sawade E, Slyman N, Dixon M, Ho L, Hoefel D, McDowall B and Newcombe G (2011) Biological filtration for the treatment of cyanobacterial metabolites, In Proceedings. *AWA OzWater Convention and Exhibition*, 9-11 May 2011, Adelaide, Australia.

5.3 Student reports

- Vache M (2012) Implementation of biological filtration for the treatment of cyanobacterial metabolites in drinking water – Biological Filtration Potential Test, Report for École Supérieure de Chimie Physique Électronique de Lyon, France.
- Lorain B (2011) Implementation of biological filtration for the treatment of cyanobacterial metabolites in drinking water – Identification of most effective substrate and temperature investigations, Report for École Supérieure de Chimie Physique Électronique de Lyon, France.

* At the time of completing final report

6 ACKNOWLEDGEMENTS

The authors would like to thank Renae Phillips, Bret Robinson, Virginie Gaget, Christopher Saint, Werner Mobius and the members of the Research and Innovation Services group, Australian Water Quality Centre, for assistance and advice throughout the project.

We gratefully acknowledge the support of the United Water/AllWater staff at Happy Valley Water Treatment Plant: Chris Copley, Jamie Sim and Neil Parker and SA Water staff at the Morgan Water Treatment Plant: John Knoblauch, Brett Hill, Matthew Seton, Andrew Prosser and Tom Stegemann. As well as our project partners, Barwon Water, Coliban Water, Hunter Water Corporation, Melbourne Water, United Water International and Veolia Water for their support and contributions to the project. We thank the Australian Water Quality Centre and the South Australian Water Corporation for supporting this project, as well as Peter Spencer and Maree Smith for their reviews.

REFERENCES

Bourne DG, Jones GJ, Blakeley RL, Jones A, Negri AP and Riddles P (1996) Enzymatic pathway for the bacterial degradation of the cyanobacterial cyclic peptide toxin microcystin LR. *Applied and Environmental Microbiology* 62(11) 4086-4094.

Bourne DG, Riddles P, Jones GJ, Smith W and Blakeley RL (2001) Characterisation of a gene cluster involved in bacterial degradation of the cyanobacterial toxin microcystin LR. *Environmental Toxicology* 16(6) 523-534.

Carlile P (1994) Further studies to investigate microcystin-LR and anatoxin-A removal from water. Report FR 0458. Marlow, Buckinghamshire, UK, Foundation for Water Research.

Carmichael WW (1992) Cyanobacteria secondary metabolites - the cyanotoxins. *Journal of Applied Bacteriology* 72(6) 445-459.

Carmichael WW (1994) The toxins of cyanobacteria. *Scientific American* 270(1) 78-86.

Chorus I and Bartram J (1999) Toxic Cyanobacteria in Water: A guide to their public health consequences, monitoring and management, World Health Organization/E&FN Spon/Routledge, London. ISBN 0-419-23930-8.

Christoffersen K, Lyck S and Winding A (2002) Microbial activity and bacterial community structure during degradation of microcystins. *Aquatic Microbial Ecology* 27(2) 125-136.

Cousins IT, Bealing DJ, James HA and Sutton A (1996) Biodegradation of microcystin-LR by indigenous mixed bacterial populations. *Water Research* 30(2) 481-485.

Craig K, Majerowski A, Bowen B, Evans GM and Stephens A (1998) Algal toxin and NOM removal using Ozone-BAC process under Australian conditions. In Proceedings. WaterTECH Conference, April 26-29, Brisbane, Australia.

Davis TW, Berry DL, Boyer GL and Gobler CJ (2009) The effects of temperature and nutrients on the growth and dynamics of toxin and non-toxic strains of *Microcystis* during cyanobacteria blooms. *Harmful Algae* 8(5) 715-725.

Elhadi SLN, Huck PM and Slawson RM (2004) Determination of System Losses of Geosmin and MIB in Bench-Scale Filtration Apparatus. *Water Quality Research Journal Canada* 39(2) 207-212.

Elhadi SLN, Huck PM and Slawson RM (2006) Factors affecting the removal of geosmin and MIB in drinking water biofilters. *Journal AWWA* 98(8) 108-119.

Falconer IR (2005) Cyanobacterial toxins of drinking water supplies: cylindrospermopsins and microcystins, CRC Press, USA. ISBN 0-415-31879-3

Fawell JK, Jart J, James HA and Parr W (1993) Blue-green algae and their toxins - analysis, toxicity, treatment and environmental control. *Water Supply - Oxford* 11 109-121.

Fleming LE, Rivero C, Burns J, Williams C, Bean J, Shea KA and Stinn J (2002) Blue green algal (cyanobacterial) toxins, surface drinking water, and liver cancer in Florida. *Harmful Algae* 1(2) 157-168.

Graham D and Hayes KP (1998) Application of solid phase microextraction for the analysis of off-flavours in water. In Proceedings. WaterTECH Conference, 26-29 April, Brisbane, Australia.

Grutzmacher G, Bottcher G, Chorus I and Bartel H (2002) Removal of microcystins by slow sand filtration. *Environmental Toxicology* 17(4) 386-394.

Hart J and Stott P (1993) Microcystin-LR removal from water. Report FR0367. Swindon, UK, Foundation for Water Research.

Hattori K (1988) Water treatment systems and technology for the removal of odor compounds. *Water Science & Technology* 20(8-9) 237-244.

Ho L (2004) The removal of cyanobacterial metabolites from drinking water using ozone and granular activated carbon. PhD Dissertation. Adelaide, Australia, University of South Australia.

Ho L, Craig K, Bustamante H and Newcombe G (2010) Removal of cyanobacterial metabolites: Laboratory evaluation of granular activated carbon. *Water: Journal of the Australian Water Association* 37(3) 65-70.

Ho L, Hoefel D, Bock F, Saint CP and Newcombe G (2007a) Biodegradation rates of 2-methylisoborneol (MIB) and geosmin through sand filters and in bioreactors. *Chemosphere* 66(11) 2210-2218.

Ho L, Hoefel D, Saint CP and Newcombe G (2007b) Isolation and identification of a novel microcystin-degrading bacterium from a biological sand filter. *Water Research* 41(20) 4685-4695.

Ho L, Meyn T, Keegan A, Hoefel D, Brookes JD, Saint CP and Newcombe G (2006) Bacterial degradation of microcystin toxins within a biologically active sand filter. *Water Research* 40(4) 768-774.

Ho L and Newcombe G (2007) Evaluating the adsorption of microcystin toxins using granular activated carbon (GAC). *Journal of Water Supply: Research and Technology* 56(5) 281-291.

Ho L and Newcombe G (2010) Granular activated carbon adsorption of 2-methylisoborneol (MIB): Pilot- and laboratory-scale evaluations. *Journal of Environmental Engineering* 136(9) 965-974.

Ho L, Sawade E and Newcombe G (2012a) Biological treatment options for cyanobacteria metabolite removal – A review. *Water Research* 46(5) 1536-1548.

Ho L, Tang T, Monis PT and Hoefel D (2012b) Biodegradation of multiple cyanobacterial metabolites in drinking water supplies. *Chemosphere* 87(10) 1149-1154.

Ho L, Tang T, Slyman N and Newcombe G (2011) Performance of GAC filters for the treatment of River Murray water. *Water: Journal of the Australian Water Association* 38(4) 65-69.

Hoefel D, Adriansen CMM, Bouyssou MAC, Saint CP, Newcombe G and Ho L (2009b) Development of an mlrA gene-directed taqMan PCR assay for quantitative assessment of microcystin-degrading bacteria within water treatment plant sand filter biofilms. *Applied and Environmental Microbiology* 75(15) 5169-5169.

Hoefel D, Ho L, Monis PT, Newcombe G and Saint CP (2009a) Biodegradation of geosmin by a novel Gram-negative bacterium; isolation, phylogenetic characterisation and degradation rate determination. *Water Research* 43(11) 2927-2935.

Hoefel D, Monis PT, Grooby WL, Andrews S and Saint CP (2005) Culture-independent techniques for rapid detection of bacteria associated with loss of chloramine residual in a drinking water system. *Applied and Environmental Microbiology* 71(11) 6479-6488.

Holst T, Jørgensena NOG, Jørgensen C and Johansen A (2003) Degradation of microcystin in sediments at oxic and anoxic, denitrifying conditions. *Water Research* 37(19) 4748-4760.

Jones GJ and Negri AP (1997) Persistence and degradation of cyanobacterial paralytic shellfish poisons (PSPs) in freshwaters. *Water Research* 31(3) 525-533.

Jones GJ and Orr PT (1994) Release and degradation of microcystin following algicide treatment of a *Microcystis aeruginosa* bloom in a recreational lake, as determined by HPLC and protein phosphatase inhibition assay *Water Research* 28(4) 871-876.

Kayal N, Newcombe G and Ho L (2008) Investigating the fate of saxitoxins in biologically active water treatment plant filters. *Environmental Toxicology* 23(6) 751-755.

Lahti K and Hiisvirta L (1989) Removal of cyanobacterial toxins in water treatment processes: review of studies conducted in Finland. *Water Supply* 7(4) 149-154.

Lam AK, Fedorak PM and Prepas EE (1995) Biotransfemation of the cyanobacterial hepatotoxin microcystin-U1, as determined by HPLC and protein phosphatase bioassay. *Environmental Science and Tecnology* 29(1) 242-246.

Magic-Knezev A and Kooij Dvd (2004) Optimisation and significance of ATP analysis for measuring active biomass in granular activated carbon filters used in water treatment. *Water Research* 38(18) 3971-3979.

McDowall B (2008) Removal of geosmin and 2-methylisoborneol from drinking water through biologically active sand filters. PhD Dissertation. Adelaide, Australia, University of Adelaide.

McDowall B, Ho L, Saint CP and Newcombe G (2007a) Biological removal of MIB and geosmin through rapid gravity filters. *Water: Journal of the Australian Water Association* 34(7) 48-54.

McDowall B, Ho L, Saint CP and Newcombe G (2007b) Removal of geosmin and 2-methylisoborneol through biologically active sand filters. *International Journal of Environment and Waste Management* 1(4) 311-320.

McDowall B, Hoefel D, Newcombe G, Saint CP and Ho L (2009) Enhancing the biofiltration of geosmin by seeding sand filter columns with a consortium of geosmin-degrading bacteria. *Water Research* 43(2) 433-440.

Metcalf JS, Beattie KA, Saker ML and Codd GA (2002) Effects of organic solvents on the high performance liquid chromatographic analysis of the cyanobacterial toxin cylindrospermopsin and its recovery from environmental eutrophic waters by solid phase extraction. *FEMS Microbiology Letters* 216 159-164.

Newcombe G (2002) Removal of Algal Toxins using Ozone and GAC, American Water Works Association, Denver, CO, USA

Newcombe G, Cook D, Brooke S, Ho L and Slyman N (2003) Treatment options for microcystin toxins: Similarities and differences between variants. *Environmental Technology* 24(3) 299-308.

Newcombe G and Nicholson B (2002) Treatment options for the saxitoxin class of cyanotoxins. *Water Science & Technology : Water Supply* 2(5/6) 271-276.

Newcombe G and Nicholson B (2004) Water treatment options for dissolved cyanotoxins. *Journal of Water Supply: Research and Technology - AQUA* 53(4) 227-239.

Nicholson BC, Rositano J and Burch MD (1994) Destruction of cyanobacterial peptide hepatotoxins by chlorine and chloramine. *Water Research* 28(6) 1297-1303.

Paerl HW, Hall NS and Calandrino ES (2011) Controlling harmful cyanobacterial blooms in a world experiencing anthropogenic and climatic-induced change. *Science of The Total Environment* 409(10) 1739-1745.

Paerl HW and Huisman J (2008) Climate: Blooms like it hot. *Science* 320 57-58.

Rapala J, Lahti K, Sivonen K and Niemalä SI (1994) Biodegradability and adsorption on lake sediments of cyanobacterial hepatotoxins and anatoxin-a. *Letters in Applied Microbiology* 19 423-428.

Saito T, Okano K, Park HD, Itayama T, Inamori Y, Neilan BA, Burns BP and Sugiura N (2003) Detection and sequencing of the microcystin LR-degrading gene, mlrA, from new bacteria isolated from Japanese. *FEMS Microbiology Letters* 229(2) 271-276.

Saitou T, Sugiura N, Itayama T, Inamori Y and Matsumura M (2002) Degradation of microcystin by biofilm in practical treatment facility. *Water Science & Technology* 46(11-12) 237-244.

Schrader K K and Blevins WT (1993) Geosmin-producing species of Streptomyces and Lyngbya from aquaculture ponds. *Canadian Journal of Microbiology* 39(9) 834-840.

Senogles P, Smith MJ and Shaw G (2002) Physical, chemical and biological methods for the degradation of the cyanobacterial toxin, cylindrospermopsin. In Proceedings. AWWA Water Quality Technology Conference & Exhibition, November 10-14, Seattle, Washington, USA, American Water Works Association.

Sherman P, Tully I and Gibson H (1995) Removal of cyanobacterial cells and toxins from drinking water with biologically active filters. In Proceedings. 16th Federal AWWA Convention, April 2-6, pp. 587-592. Sydney, Australia.

Smith MJ, Shaw GR, Eaglesham GK, Ho L and Brookes JD (2008) Elucidating the factors influencing the biodegradation of cylindrospermopsin in drinking water sources. *Environmental Toxicology* 23(3) 413-421.

Tsuji K, Setsuda S, Watanuki T, Kondo F, Nakazawa H, Suzuki M and Harada K (1996) Microcystin levels during 1992-95 for Lakes Sagami and Tsukui-Japan. *Natural Toxins* 4(4) 189-194.

UKWIR (1996). Pilot scale GAC tests to evaluate toxin removal. Report No. 96/DW/07/1. London, UK, UK Water Industry Research Ltd.

Velten S, Hammes F, Boller M and Egli T (2007) Rapid and direct estimation of active biomass on granular activated carbon through adenosine tri-phosphate (ATP) determination *Water Research* 41(9) 1973-1983.

Wang H, Ho L, Lewis D M, Brookes JD and Newcombe G (2007) Discriminating and assessing adsorption and biodegradation removal mechanisms during granular activated carbon filtration of microcystin toxins. *Water Research* 41(18) 4262-4270.

Wang H, Lewis D, Newcombe G, Brookes J and Ho L (2009) Separated adsorption and bacterial degradation of microcystins in GAC filtration. *International Journal of Environment and Waste Management* 3(3/4) 236-243.

Watanabe MF, Tsuji K, Watanabe Y, Harada KI and Suzuki M (1992) Release of heptapeptide toxin (microcystin) during the decomposition process of Microcystis aeruginosa. *Natural Toxins* 1 48-53.

APPENDIX I

BIOLOGICAL FILTRATION POTENTIAL (BFP) TEST PROCEDURE

Scope

This method is used to establish the potential of full- or pilot-scale water treatment plant (WTP) filters to function as biological filters for the removal of cyanobacterial taste and odour compounds and toxins (cyanobacterial metabolites). The test utilises laboratory-scale filter columns, batch testing, and several tests to quantify and characterise the biomass and biological activity on the filter. The results of the test are indicative of the current biological activity of the filter and the potential of the filter to develop a biomass capable of reducing or eliminating cyanobacterial metabolites. The media to be tested can be sand, filter coal/anthracite, granular activated carbon (GAC) or any medium in use at the plant. It can be used with a range of water quality.

Principle

Water from the treatment plant is sampled from a point just prior to the full-scale filters (test water), and the filter medium is sampled from the top of the filter, or from the top of the appropriate layer in a multimedia filter.

This test is divided into several sections:

Batch testing:

Cyanobacterial metabolites are spiked into the test water which is then sampled over a period of two weeks to determine if the water contains organisms capable of degrading the metabolites. A sterilised control is also tested to determine any other loss of the metabolites in the system through adsorption, volatilisation, or chemical degradation.

Laboratory-scale filter column testing:

Laboratory-scale filter columns have been shown to successfully simulate full-scale performance for the removal of cyanobacterial metabolites (Ho 2004; Ho et al. 2010; Ho and Newcombe 2010). Test water spiked with metabolites is pumped through a glass column packed with the filter medium to be tested. The flow rate is adjusted to maintain an empty bed contact time (EBCT) of 15 minutes. The water is sampled prior to filtration through the column and afterwards, and the removal through filtration is calculated. A second column is run in parallel containing the same filtration medium after a sterilisation procedure. The difference in the removal of the metabolites can be attributed to biological degradation.

Biofilm characterisation and biological activity quantification:

Samples of the filter medium are extracted and characterised using three techniques:

Adenosine triphosphate (ATP): This provides an indication of the overall level of biofilm activity and viability. A highly active biofilm is likely to be more efficient in removing the cyanobacterial metabolites of concern compared with biofilm of lower activity. ATP testing will also confirm that filters within a WTP are biologically active, as inadvertent operational changes within a WTP can sometimes result in destruction of filter biofilm (e.g. pre-chlorination).

mlrA gene abundance:

This gene is known to be responsible for the initial step in the degradation of microcystin toxins by microcystin degrading bacteria. Detection of the mlrA gene in the biofilm indicates only that microcystin degrading bacteria are present, and there is a potential for the filter to degrade the toxins. mlrA gene quantification provides an indication of the abundance of microcystin degrading bacteria within the biofilm of a filter, where the efficiency of removals has been directly linked to the abundance of these bacteria.

16S rRNA abundance:

This parameter provides an indication of total bacterial abundance within a filter biofilm and can be used to assess the overall state of the biofilm especially following seasonal and/or operational changes within the WTP. This parameter is used along with the ATP data to provide an overall status of the filter biofilm.

Sampling

Water

200 L of sample water is required to complete these tests. Water should be sampled at the point just prior to the filter to be tested. Storage below 4°C will preserve sample integrity until analysis.

Filter medium

The filter medium should be sampled after backwash from the top of the filter, or from the top of the appropriate layer if it is a multimedia filter. 500 mL of filter medium is sufficient for all tests. The sample should be kept wet (with the sample water) and transported for testing as soon as is practicable.

Procedures

Batch test procedure

Two 5 L bioreactors are used for the batch test. A volume of 9 L of the test water is spiked with a mixture of the metabolites. Table 9 shows the metabolites and the concentrations used. 4.5 L of the spiked water is placed in each vessel. One bioreactor is then autoclaved to inactivate bacteria - this is the batch test control. A 500 mL sample is taken from each vessel to determine exact initial concentrations of the metabolites. Samples are then taken on days 2, 4, 6, 8, 10, 12, 14 to determine concentration of the metabolites. Aseptic techniques are used for sampling to ensure there is no contamination of the test water.

Table 9: Batch test metabolite concentrations

Metabolite	**Concentration**
MIB	50 ng/L
Geosmin	50 ng/L
Microcystins LR and LA	Total 20 µg/L
Saxitoxins	Total 10 µg/L
Cylindrospermopsin	10 µg/L

Laboratory-scale filter test

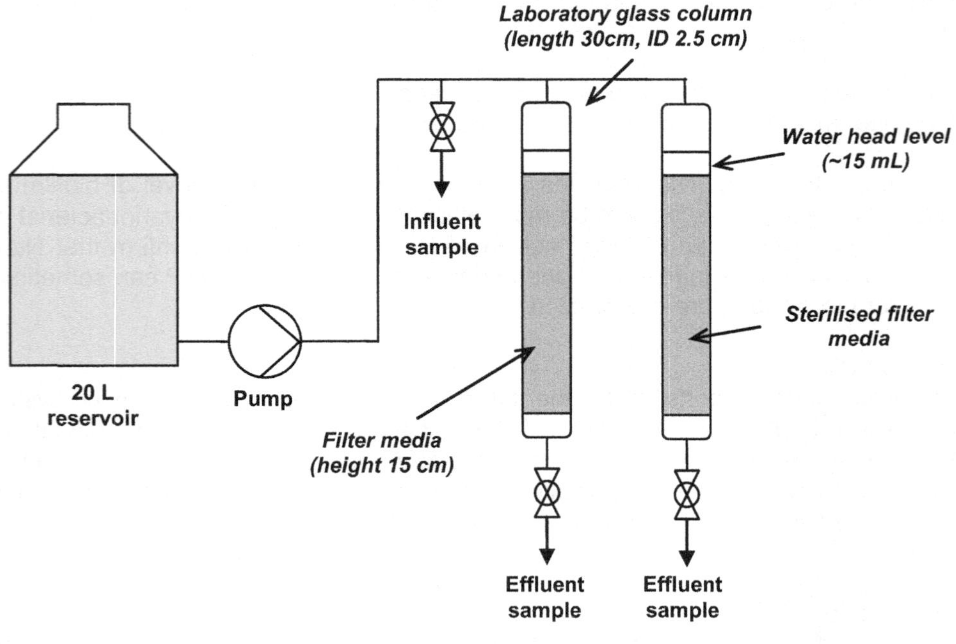

Figure 56: Laboratory-scale filter testing apparatus

Procedure

i) Measure 75 mL of wet filter medium into a measuring cylinder, tap sides to remove air bubbles
ii) Rinse filter medium into glass column filter (with tap closed) with sample water, stopper column and invert several times to remove air bubbles. Ensure filter medium is fully immersed, with approximately 3 cm of sample water above the surface of the filter medium
iii) Repeat i), autoclave filter medium sample to sterilise
iv) Place sterilised filter medium sample in second column as described in ii)
v) Place sample water in the reservoir, dose with the appropriate concentration of metabolites
vi) Start pump and adjust flow rate to 5 mL per minute (EBCT=15 minutes)
vii) Allow to run for 1 hour to stabilise
viii) Take sample at inlet sample port of the appropriate volume for the required analysis
ix) Take sample at the outlet of each column
x) Repeat sampling at 2 day intervals (or timing as appropriate) for 12 days

Biofilm analysis

Adenosine triphosphate (ATP)

The ATP analysis was adapted from the method provided with the ATPlite luminescence ATP detection assay system (Perkin Elmer, USA) to allow direct analysis of filter medium samples. Filter medium samples are rinsed with ultrapure water prior to analysis to remove any excess suspended particles. Samples are analysed in a 96 well isoplate using the Wallac 1420 multilabel counter (Perkin Elmer, USA). The luminescence outputs are compared to ATP standard curves. Final ATP concentrations are multiplied by 4 to account for the use of the 200 µL of lysis solution used for the total weight of sand. This result is then divided by the wet weight of the sand to obtain mol/L ATP/g filter medium. This result is then multiplied by the density of the filter medium and the final result is reported as mol/L ATP per mL filter material. Analysis is conducted in triplicate and the average taken. Details are given in McDowall, (2008).

mlrA gene and 16S rRNA abundance

Procedures for mlrA gene and 16S rRNA abundance can be found in Hoefel et al. (2009b) and Hoefel et al. (2005), respectively.

<u>Reporting</u>

Batch tests are plotted as concentration or fraction remaining vs time. Results of column tests can be plotted with influent and effluent values on the same graph, or as percent removals or fraction remaining, of a particular metabolite.

<u>Interpretation</u>

These three tests provide information about:

1. The ability of the full-scale filters to biologically remove a range of cyanobacterial metabolites at the time of sampling. This will allow the water authority to plan the level of alternative treatment processes that may be necessary if the plant is challenged with any of these metabolites. Or, in the case of GAC, it will allow the authority to determine whether the medium requires replacement prior to the next cyanobacterial challenge
2. The existence of degrading organisms in the inlet water to the filters, and thus the potential to develop a biofilm capable of degrading the metabolites. If 1, above, indicates that the filter is not currently capable of degrading metabolites, this may provide a justification for modifications to filter media, or operational parameters to encourage further biofilm development with the aim of developing a degrading biofilm.
3. The health and extent of the biofilm, and the presence or absence of organisms capable of degrading microcystins. With information provided 1 and 2, above, this will allow the water authority to make informed decisions regarding any operational changes that may promote the growth of a degrading biofilm.

APPENDIX II

Additional results from BFP test

Barwon Water, Wurdee Boluc Water Treatment Plant

Laboratory-scale filtration

Figure 57 shows the results of the column testing to identify biological removal of MIB, geosmin, STXeq and mLR during filtration through sterile and non-sterile anthracite from the Wurdee Boluc WTP. The influent water was sampled from the settled water ducts at the plant.

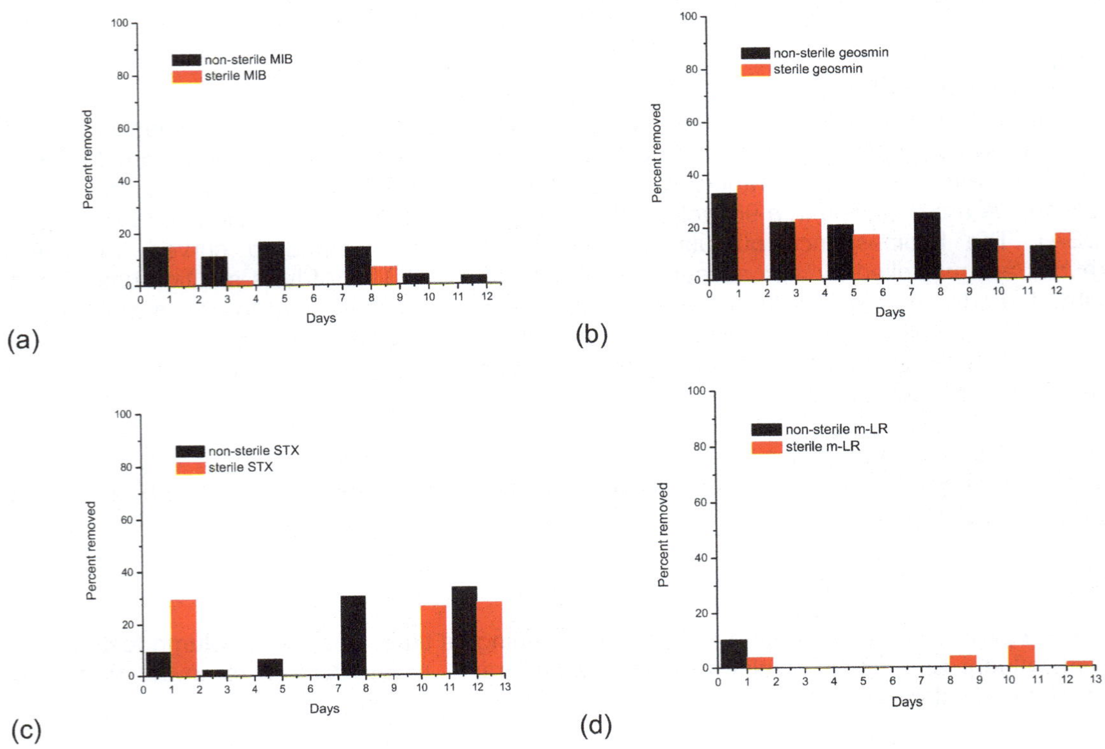

Figure 57: Column experiments, percent removals over 12 days: MIB, initial concentration 46 ng/L (a); geosmin, initial concentration 39 ng/L (b); saxitoxin equivalents, initial concentration 2.7 µg/L (c); microcystin LR, initial concentration 6.2 µg/L (d)

The results suggest no measurable biological removal of these metabolites during laboratory column filtration.

Batch experiments

Settled water from Wurdee Boluc WTP was spiked with MIB, geosmin, STXeq, mLR and mLA and cylindrospermopsin. Another water sample was sterilised prior to spiking. The concentrations of the metabolites sampled over 15 days are shown in Figure 58. Any decrease in concentration of the non-sterile sample indicates the presence of degrading organisms in the influent water to the filters at the full scale, and the potential of these organisms to colonise subsequent filtration media.

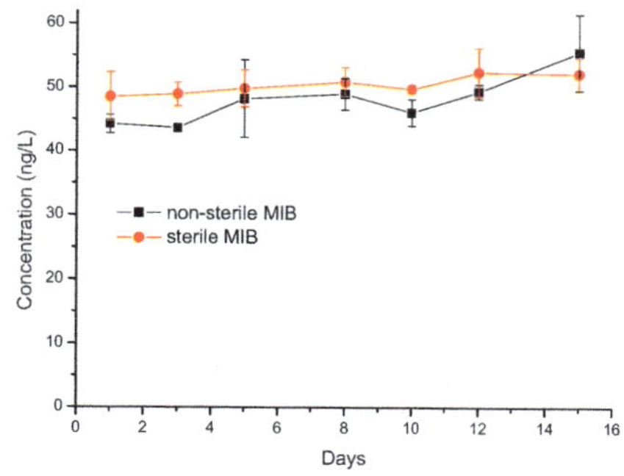

(a)

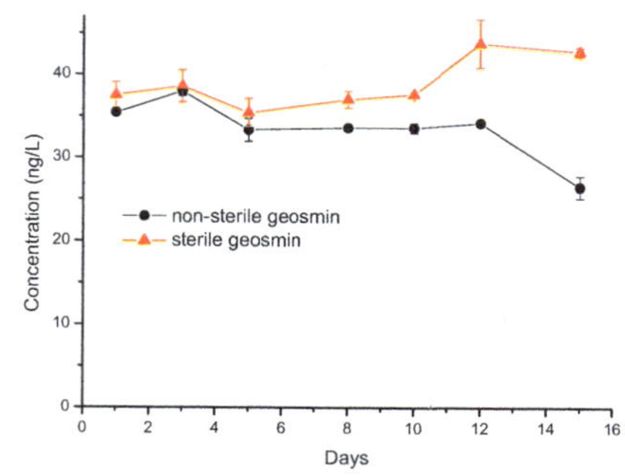

(b)

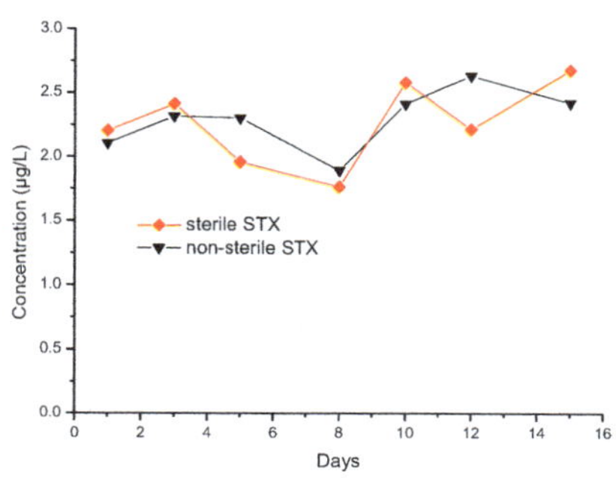

(c)

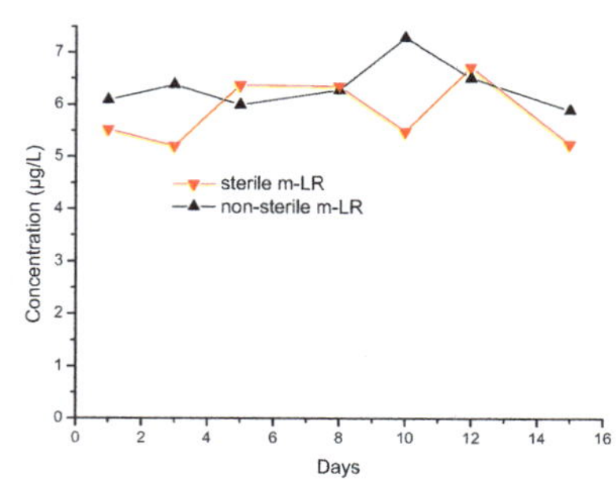

(d)

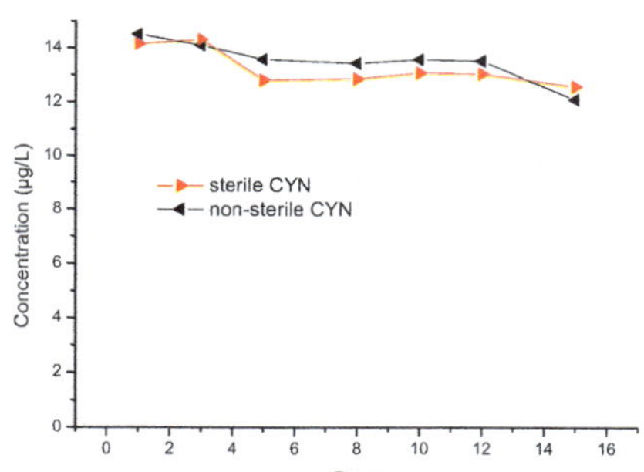

(e)

Figure 58: Batch experiments in settled water, concentration as a function of time: MIB (a); geosmin (b); STXeq (c); microcystin LR (d); cylindrospermopsin (e).

No degradation of the metabolites occurred in the batch experiments in the time period investigated, indicating the absence of degrading organisms in the influent to the filters at this WTP.

Summary

The filters at Wurdee Boluc WTP do not display any biological removal of metabolites, and the influent water did not contain degrading bacteria that could colonise the filters at the time of sampling. The potential of biological filtration at Wurdee Boluc WTP under the current conditions appears low.

Melbourne Water, Winneke Water Treatment Plant

Laboratory-scale filtration

Figure 59 shows the results of the column testing to identify biological removal of MIB, geosmin, STXeq and mLR during filtration through sterile and non-sterile sand from the Winneke WTP. The influent water was sampled from the settled water ducts at the plant.

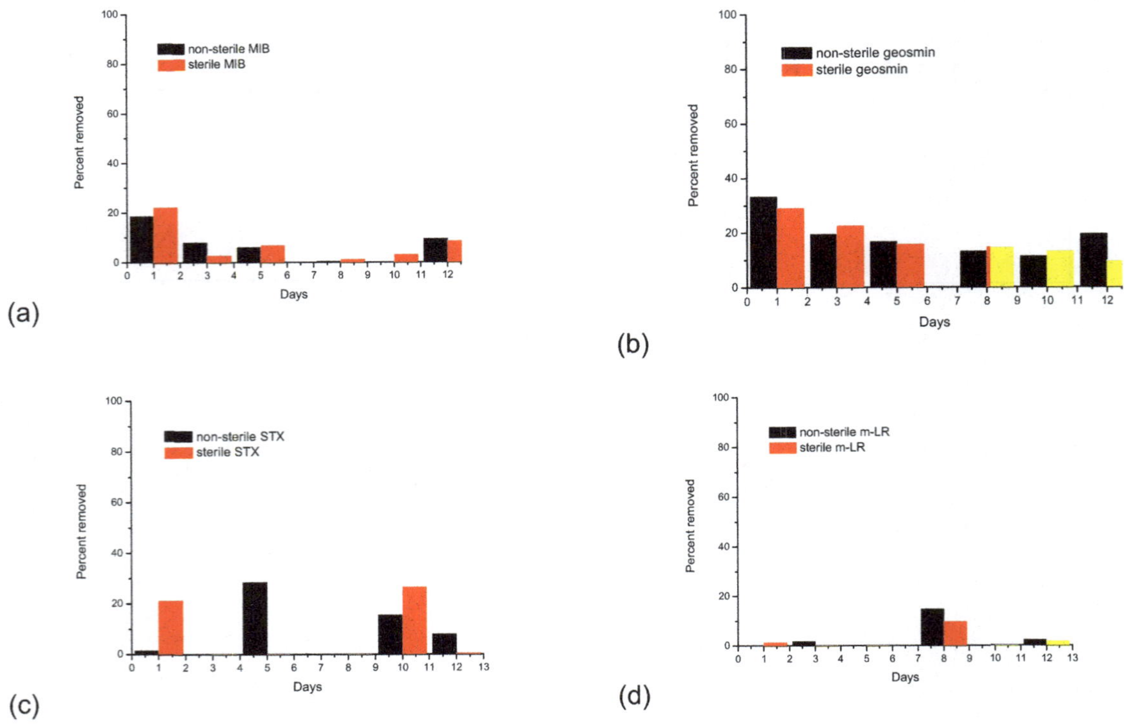

Figure 59: Column experiments, percent removals over 12 days: MIB, initial concentration 57 ng/L (a); geosmin, initial concentration 33 ng/L (b); saxitoxin equivalents, initial concentration 2.9 µg/L (c); microcystin LR, initial concentration 5.4 µg/L (d)

The results suggest no measurable biological removal of these metabolites was evident during laboratory column filtration investigation.

Batch experiments

Settled water from Winneke WTP was spiked with MIB, geosmin, STXeq, mLR and mLA and cylindrospermopsin. Another water sample was sterilised prior to spiking. The concentrations of the metabolites sampled over 15 days are shown in Figure 60. Any decrease in concentration of the non-sterile sample indicates the presence of degrading organisms in the influent water to the filters at the full scale, and the potential of these organisms to colonise subsequent filtration media.

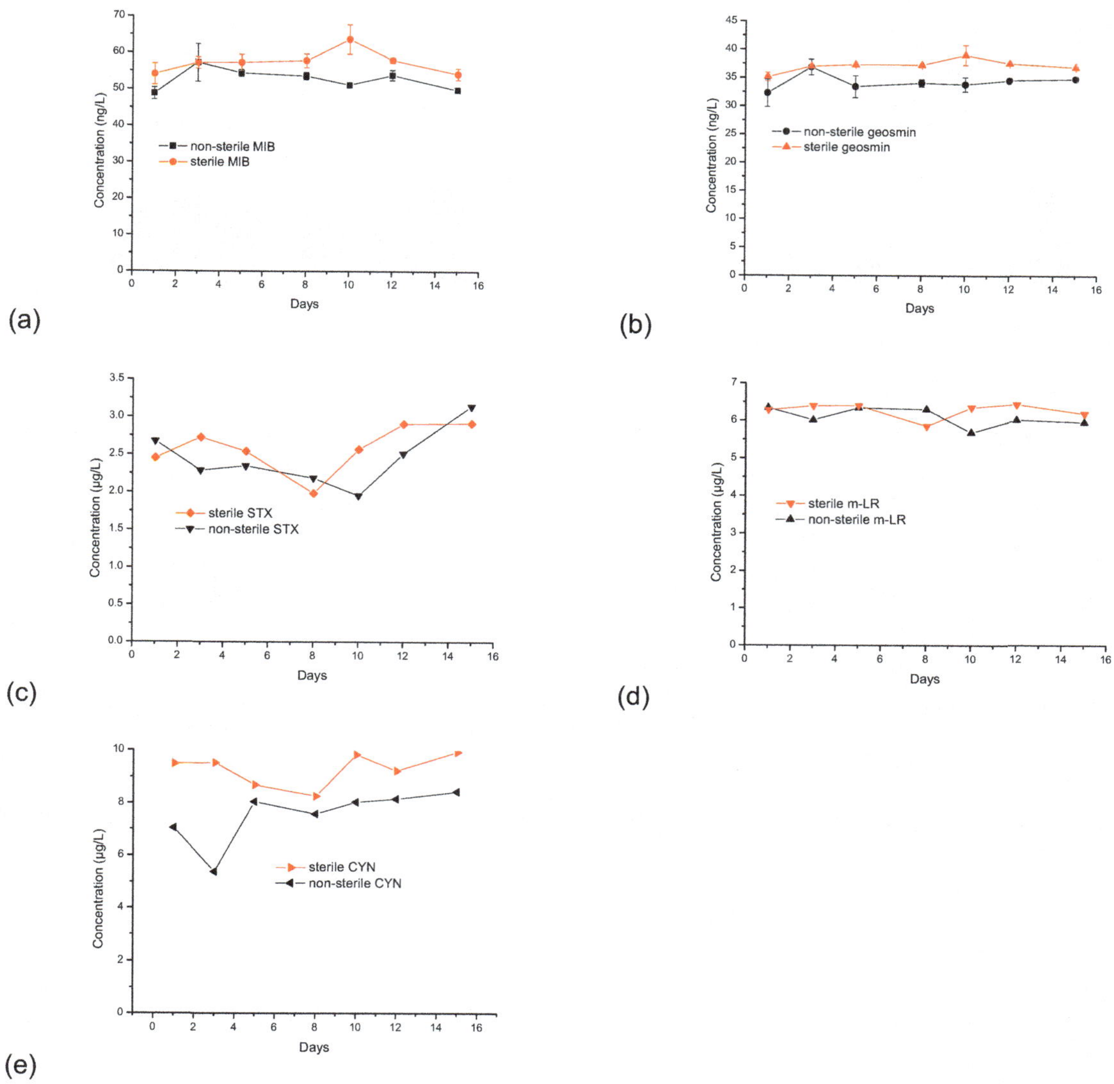

Figure 60: Batch experiments in settled water, concentration as a function of time: MIB (a); geosmin (b); STXeq (c); microcystin LR (d); cylindrospermopsin (e).

No degradation of the metabolites occurred in the batch experiments, indicating the absence of degrading organisms in the influent to the filters at this WTP.

Summary

The filters at Winneke WTP do not display any biological removal of metabolites, and the influent water did not contain degrading bacteria that could colonise the filters at the time of sampling. The potential of biological filtration at Winneke WTP under the current operating conditions appears low.

Veolia Water, Illawarra Water Treatment Plant

Laboratory-scale filtration

Figure 61 shows the results of the column testing to identify biological removal of MIB, geosmin, STXeq and mLR during filtration through sterile and non-sterile anthracite from the Illawarra WTP. The influent water was sampled from the settled water ducts at the plant.

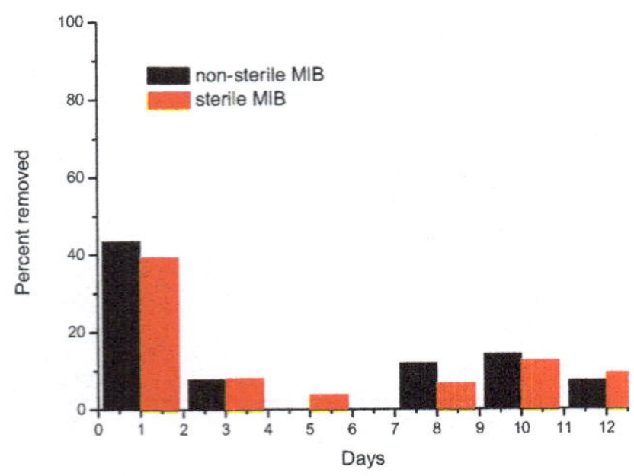

(a)

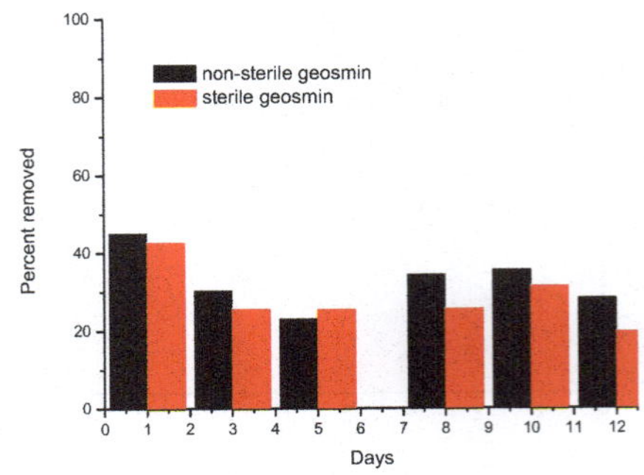

(b)

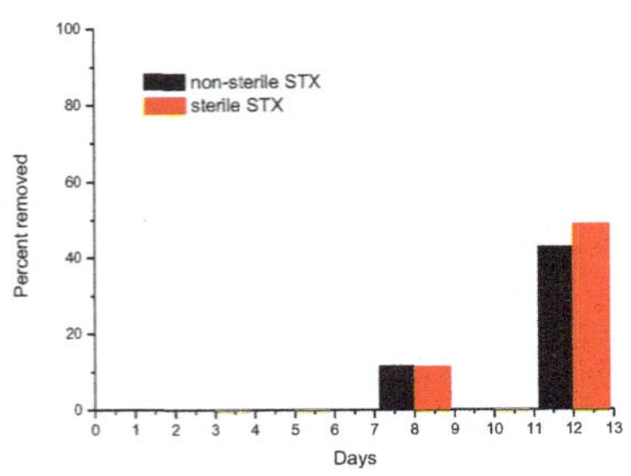

(c)

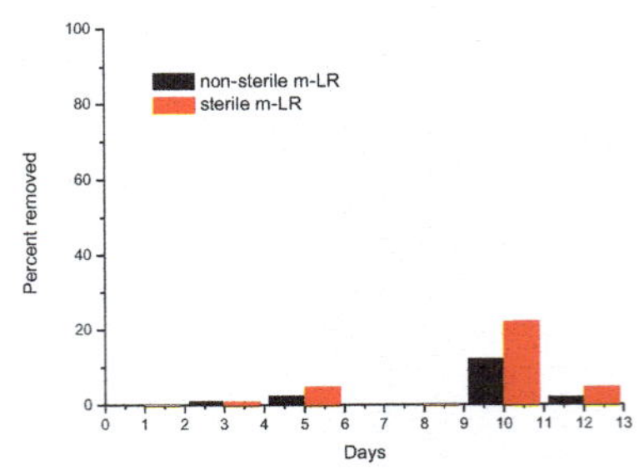

(d)

Figure 61: Column experiments, percent removals over 12 days: MIB, initial concentration 46 ng/L (a); geosmin, initial concentration 36 ng/L (b); saxitoxin equivalents, initial concentration 0.6 µg/L (c); microcystin LR, initial concentration 5.6 µg/L (d)

The results suggest no measurable biological removal of these metabolites during laboratory column filtration.

Batch experiments

Settled water from Illawarra WTP was spiked with MIB, geosmin, STXeq, mLR and mLA and cylindrospermopsin. Another water sample was sterilised prior to spiking. The concentrations of the metabolites sampled over 15 days are shown in Figure 62. Any decrease in concentration of the non-sterile sample indicates the presence of degrading organisms in the influent water to the filters at the full scale, and the potential of these organisms to colonise subsequent filtration media.

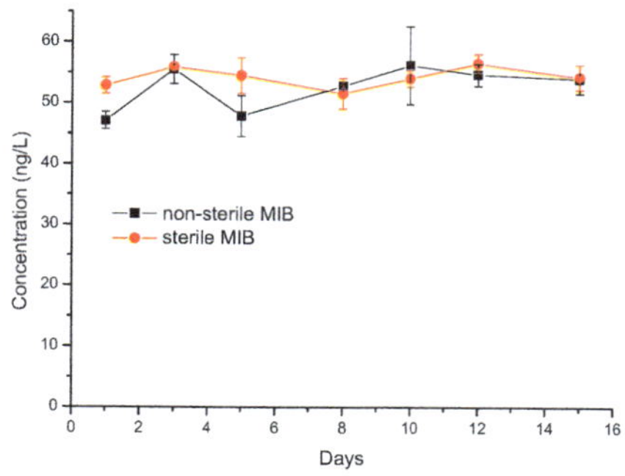

(a)

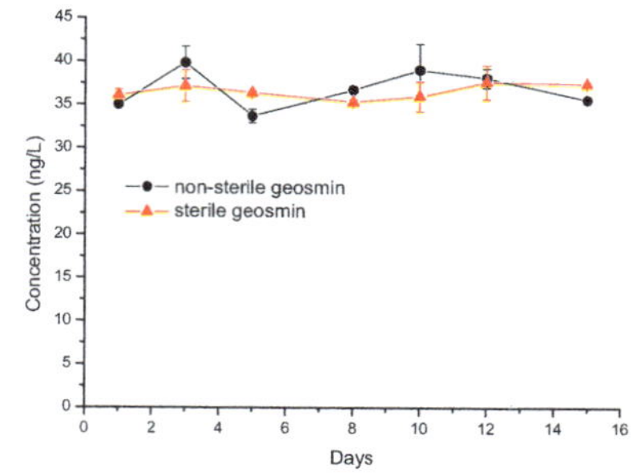

(b)

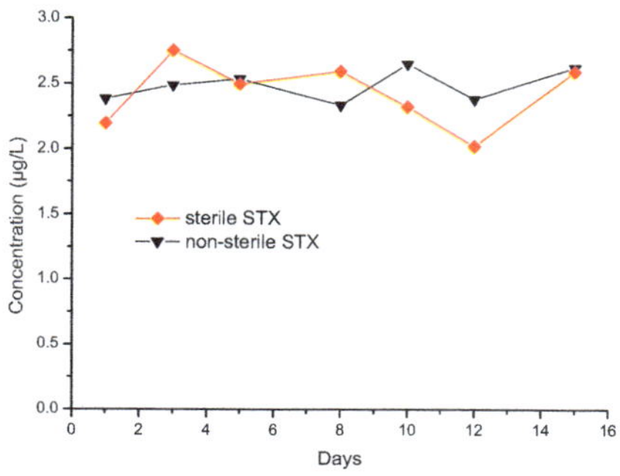

(c)

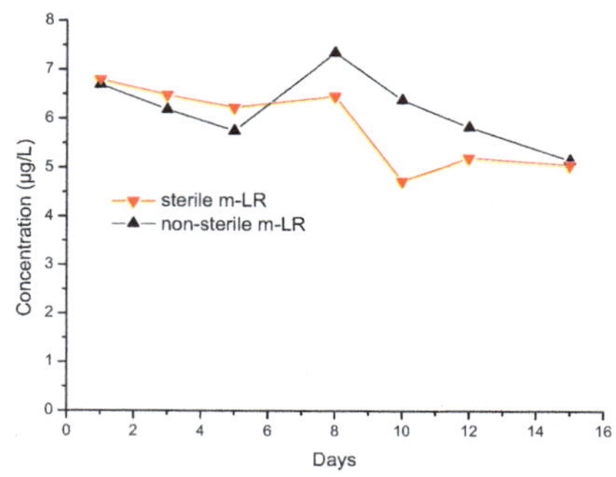

(d)

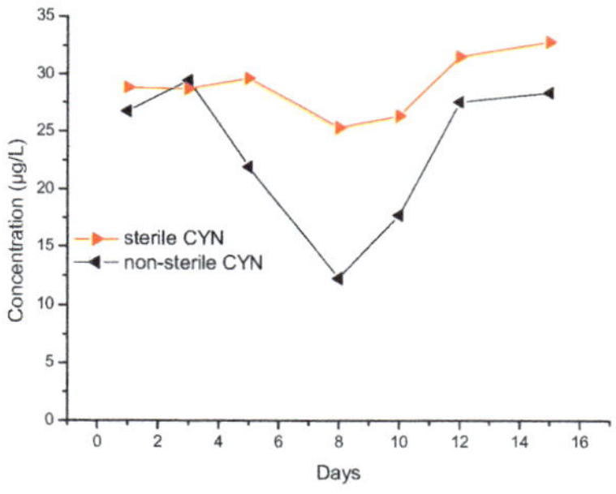

(e)

Figure 62: Batch experiments in settled water, concentration as a function of time: MIB (a); geosmin (b); STXeq (c); microcystin LR (d); cylindrospermopsin (e).

No degradation of the metabolites occurred in the batch experiments, indicating the absence of degrading organisms in the influent to the filters at this WTP.

Summary

The filters at Illawarra WTP do not display any biological removal of metabolites, and the influent water did not contain degrading bacteria that could colonise the filters at the time of sampling. The potential of biological filtration at Illawarra WTP under the current operating conditions appears low.

IWA Publishing's authorised EU representative for General Product
Safety Regulations is Diane D'Arras, 15 rue Duret, 75116 Paris,
France, e-mail: safety@iwap.co.uk.

Printed and bound by CPI Group (UK) Ltd, Croydon, CR0 4YY
17/09/2026
02218814-0001